An Outline of
Set Theory

An Outline of Set Theory

James M. Henle
Department of Mathematics
Smith College

Dover Publications, Inc.
Mineola, New York

Copyright

Copyright © 1986 by James M. Henle
All rights reserved.

Bibliographical Note

This Dover edition, first published in 2007, is a corrected and unabridged republication of the work first published as part of the Problem Books in Mathematics Series by Springer-Verlag Inc., New York, in 1986.

Library of Congress Cataloging-in-Publication Data

Henle, James M.
    An outline of set theory / James M. Henle.
    p. cm.
    Originally published: New York : Springer-Verlag, c1986, in series: Problem books in mathematics.
    ISBN-13: 978-0-486-45337-8
    ISBN-10: 0-486-45337-5
    1. Set theory. 2. Logic, Symbolic and mathematical.

QA248 .H43 2007
511.3'22—dc22

2006050271

Manufactured in the United States of America
Dover Publications, Inc., 31 East 2nd Street, Mineola, N.Y. 11501

Preface

This book is designed for use in a one semester problem-oriented course in undergraduate set theory. The combination of level and format is somewhat unusual and deserves an explanation.

Normally, problem courses are offered to graduate students or selected undergraduates. I have found, however, that the experience is equally valuable to ordinary mathematics majors. I use a recent modification of R. L. Moore's famous method developed in recent years by D. W. Cohen [1]. Briefly, in this new approach, projects are assigned to groups of students each week. With all the necessary assistance from the instructor, the groups complete their projects, carefully write a short paper for their classmates, and then, in the single weekly class meeting, lecture on their results. While the emphasis is on the student, the instructor is available at every stage to assure success in the research, to explain and critique mathematical prose, and to coach the groups in clear mathematical presentation.

The subject matter of set theory is peculiarly appropriate to this style of course. For much of the book the objects of study are familiar and while the theorems are significant and often deep, it is the methods and ideas that are most important. The necessity of reasoning about numbers and sets forces students to come to grips with the nature of proof, logic, and mathematics. In their research they experience the same dilemmas and uncertainties that faced the pioneers. They will, for example, discover in some chapters that deeper results in earlier chapters are necessary before work can proceed.

Students do not always solve the problems completely on their own. They do, however, learn what proofs are and how to organize and write them, and while lectures on this material might easily bore, students find the experience of doing it themselves exciting and rewarding. It is familiar enough to be reassuring and different enough to be challenging.

More set theory is included here than one can reasonably use. I cover roughly 35 to 40 projects in a semester. The last three chapters are independent of each other and can be used selectively or omitted. Sections of other chapters may also be skipped or summarized, particularly the last few in Chapter 7.

I am indebted first of all to David Cohen, for the example of his outstanding teaching, and to my students for their intelligence and unflagging good humor. I only hope that my confidence in this approach is not based entirely on a teacher who might succeed with *any* method, and students who might prevail under *any* regimen. I greatly appreciate the support of Smith College and the encouragement of its most collegial mathematics department. Thanks are also due to Marcia Groszek for the Tennyson quotation, and special thanks to Carlos Di Prisco for his very timely suggestions and advice.

References

[1] D. W. Cohen, "A Modified Moore Method for Teaching Undergraduate Mathematics," *Am. Math. Monthly* **89**, no. 7, 1982.

J. M. Henle

Contents

Preface ... v

Introduction ... 1

Part One Projects

1. Logic and Set Theory ... 7
2. The Natural Numbers ... 15
3. The Integers ... 21
4. The Rationals ... 23
5. The Real Numbers ... 25
6. The Ordinals ... 27
7. The Cardinals ... 33
8. The Universe ... 37
9. Choice and Infinitesimals ... 41
10. Goodstein's Theorem ... 45

Part Two Suggestions

1. Logic and Set Theory ... 51
2. The Natural Numbers ... 55
3. The Integers ... 57
4. The Rationals ... 59
5. The Real Numbers ... 63
6. The Ordinals ... 67
7. The Cardinals ... 71

8.	The Universe	81
9.	Choice and Infinitesimals	85
10.	Goodstein's Theorem	91

Part Three Solutions

1.	Logic and Set Theory	97
2.	The Natural Numbers	101
3.	The Integers	105
4.	The Rationals	109
5.	The Real Numbers	115
6.	The Ordinals	119
7.	The Cardinals	123
8.	The Universe	129
9.	Choice and Infinitesimals	133
10.	Goodstein's Theorem	137

Index 141

Introduction

As a branch of mathematics, set theory is less than one hundred years old, yet it occupies a unique and critical position. Set-theoretic principles and methods pervade mathematics. Set-theoretic results have shaken the worlds of analysis, algebra, and topology. Simple questions about sets have split the mathematical community into hostile camps, and the romance of its infinite sets have charmed and challenged philosophers as nothing else in mathematics.

A Little History

Mathematics is a living creature, growing as occasions demand and circumstances permit. Every now and then it must pause to organize and reflect on what it is and where it comes from. This happened in the third century B.C. when Euclid thought he had derived most of the mathematical results known at the time from five postulates. By the end of the nineteenth century, it was ready to happen again. Methods and structures had long outstripped Euclid and the need arose for a clearer understanding of number, proof, and existence.

In searching for underlying principles, mathematicians were led naturally to sets. It was discovered in the early twentieth century that virtually the entire body of mathematics could be described in terms of sets. More importantly, a host of critical questions had surfaced and many of these were intimately connected with sets.

Sets seemed basic and uncomplicated. The task of putting mathematics on a firm foundation appeared to be simply a matter of reducing it to sets. Unfortunately the naive approach to set theory led to trouble.

The most stunning example of the difficulty was discovered by Bertrand Russell in 1901. It was thought that any describable collection of objects must be a set. This being so, let R be the collection of all sets which are not members of themselves. We will use the notation $x \in y$ to mean "x is a member of y," and $x \notin y$ to mean "x is not a member of y," and so we write:

$$R = \{x \mid x \notin x\}.$$

Is $R \in R$? If $R \in R$ then by the definition of R, R is not a member of itself, i.e., $R \notin R$. Conversely, if $R \notin R$, then R is one of those sets which is not a member of itself, hence it is one of the sets in R, or $R \in R$! We have a contradiction we cannot escape—unless R is not a set.

Over the decades following the discovery of such problems, a collection of first principles or axioms was formed which appeared (and still appears) to avoid paradoxes. The system is called Zermelo–Fraenkel set theory or ZF after its originators, Ernst Zermelo and Abraham Fraenkel. In addition to occupying a strategic location in mathematics, ZF is studied for itself by a growing number of mathematicans. The father of modern set theory was Georg Cantor.

A Little Philosophy

The exploration of set theory revealed many divisions in mathematics. To choose an axiom system is in some sense to legislate mathematical truth. Not all mathematicans agreed to the choices. Disputes were frequent and sometimes bitter, particularly the ones involving the existence of infinite sets and the use of infinite methods.

While some mathematicans (platonists) felt the axioms were either true or false, yet others (formalists) felt that truth was a relative notion, that absolute truth did not exist. They asserted that the axioms were like the rules of a game. The rules in this case were well-chosen because they allowed players to investigate all of mathematics.

As philosophers still debate the appropriateness of set theory in general and ZF in particular as a basis for mathematics, yet another group has begun questioning the very need for a foundation. Mathematics is not in need of definition, it is claimed, mathematics simply exists. Mathematics is what our contemporary civilization conceives it to be. Its nature changes as it grows and reflects our culture.

Our Intentions

We intend to present Zermelo–Fraenkel set theory and show how it avoids Russell's paradox. Following this, we will use sets to construct a good deal of mathematics, specifically the natural numbers, the integers, the rational numbers, and the reals.

We will examine the mathematics of infinity in its many forms. In the course of this, we will consider some of the questions that have plagued and divided mathematicians in this century. We will conclude with some set theory for its own sake: some new axioms, some big sets, an application of infinity to finite sets, a discussion of the celebrated Axiom of Choice, and an introduction to nonstandard analysis.

Our Method

The format of this book also deserves some explanation. The outlines of set theory are presented here in Part One as 44 projects. These projects, in effect, ask the reader to function as mathematician. Usually theorems and definitions are provided and the proofs are not. Occasionally even definitions are omitted! Some projects are routine; others are quite challenging. Part Two contains extensive suggestions and comments. Complete (though terse) solutions are provided in Part Three.

Why are we doing this to you?

We believe there is no better way to learn mathematics than to do it: to search for it, to find it, to organize it, and to write it. We are confident of the reader's ability to do much of the work, possibly with the assistance of an instructor. We are also optimistic that the reader will enjoy the challenge.

Our Biases

We hope our biases don't show. While mathematicians disagree on the role of set theory, most agree that the construction of number systems from sets is a significant act, one which tells us a great deal about mathematics. There is also general agreement about the importance of set-theoretic ideas in mathematics.

Platonists will not be disturbed by extravagant claims of truth for ZF or other axiom systems. Formalists will find satisfaction in our rigorous adherence to ZF. Modern philosophers may approve too, for our procedure mirrors the organic growth of mathematics. While some projects assign theorems for the reader to prove, others ask for theorems to be discovered, structures to be defined. You will find occasionally that to complete one project, say a theorem on rational numbers, it is necessary to retrace your steps and prove a theorem about integers, natural numbers, or sets, that had not seemed important before. All of this is very much the way mathematics has in fact developed.

PART ONE
PROJECTS

CHAPTER 1
Logic and Set Theory

> The most distinct and beautiful statements of any truth must take at last the mathematical form. We might so simplify the rules of moral philosophy, as well as of arithmetic, that one formula would express them both.
>
> H. D. Thoreau

The goal in this first section is to become acquainted with logical notation and most of the axioms of Zermelo–Fraenkel set theory. We will also show how a few very important mathematical objects such as functions and relations can be formed from sets. Just as we have chosen to build mathematics using set theory, we will build set theory using logic.

The Language of Set Theory, $\mathscr{L}$

Basic symbols:

$=$	(equals)
$\in$	(is an element of)
$a, b, c, \ldots, z$	(variables)
$\neg$	(it is not true that)

∧	(and)
∨	(or)
→	(implies)
↔	(if and only if)
∀	(for all)
∃	(there exists)
(,)	(parentheses).

These symbols are used to form statements or formulas as follows:

(1) Basic statements of the form: $a = b$ or $a \in b$ (where a and b can be replaced by any other variables).
(2) Compound statements: if φ and ψ are statements, then
   (a) $\neg \varphi$ is also a statement,
   (b) $(\varphi \wedge \psi)$
   (c) $(\varphi \vee \psi)$
   (d) $(\varphi \rightarrow \psi)$ and
   (e) $(\varphi \leftrightarrow \psi)$ are also statements.
   If $\varphi(x)$ is a statement which says something about the variable x, then
   (f) $\forall x(\varphi(x))$ and
   (g) $\exists x(\varphi(x))$ are statements (where x can be replaced by any variable).
   We will occasionally write $\forall x \varphi(x)$ or $\exists x \varphi(x)$ (omitting some parentheses) when the meaning is clear.

These statements are interpreted as follows:

$\neg \varphi$	is true iff φ is false "iff" is an (English) abbreviation for "if and only if".
$\varphi \wedge \psi$	is true iff both φ and ψ are true.
$\varphi \vee \psi$	is true iff either φ or ψ is true (or both are true).
$\varphi \rightarrow \psi$	is true iff either ψ is true or φ is false.
$\varphi \leftrightarrow \psi$	is true iff either φ and ψ are both true, or else both are false.
$\forall x \varphi(x)$	is true iff the statement $\varphi(x)$ is true about all sets.
$\exists x \varphi(x)$	is true iff the statement $\varphi(x)$ is true about at least one set.

Examples:

To say that b is a member of both d and q we could write:

$$(b \in d \wedge b \in q).$$

1. Logic and Set Theory

To say that p and h are different sets:
$$\neg p = h.$$

To say that r has no members:
$$\forall y \, \neg (y \in r).$$

To say that u is a subset of t:
$$\forall j (j \in u \to j \in t).$$

To say that z has exactly one member:
$$\exists i \forall m (m \in z \leftrightarrow m = i).$$

PROJECT #1. For the purposes of this exercise, assume that the only sets which exist in the entire universe are the following: $a = \{b, c\}$, $b = \{\ \}$, $c = \{e\}$, $d = \{c, e\}$, $e = \{b\}$. We mean by this notation that the only members of a are b and c, that b has no members, etc.

1. Which of the following statements are true?

   (i) $b \in e$ (ii) $e \in b$ (iii) $b \in d$ (iv) $(a \in c \lor c \in d)$
   (v) $(b \in c \to a \in a)$ (vi) $(e \in d \leftrightarrow d \in e)$ (vii) $(e \in c \land a \in c)$
   (viii) $\forall x \, \neg (x \in b)$ (ix) $\forall q (q \in c \to q \in a)$ (x) $\forall n (n \in e \to n \in a)$
   (xi) $\exists k (k \in d)$ (xii) $\forall s \exists t (s \in t)$ (xiii) $\forall s \exists t (t \in s)$
   (xiv) $\exists h \forall i ((i \in e \to i \in h) \land \neg h = e)$
   (xv) $\exists g \exists w ((g \in e \land w \in e) \land \neg g = w)$

(2) Explain why these last three are statements according to the rules (1) and (2) above.

Notation. To enrich our language $\mathscr{L}$, we can add additional symbols: $\cap, \cup, \subseteq, \subsetneq, \neq, \notin, \{\ ,\ \}$.

$(a \cap b)$	stands for the intersection of the sets a and b, that is, the collection of all elements in both a and b.
$(c \cup d)$	stands for the union of the sets c and d, that is, the collection of all elements in either c or d.
$(e \setminus f)$	stands for the collection of all elements of e not in f.
$(g \subseteq h)$	means that g is a subset of h, that is, that all elements of g are also in h.
$(i \subsetneq j)$	means that i is a proper subset of $j (i \subseteq j$ but $\neg i = j)$.
$(k \neq l)$	means $\neg k = l$.

$(m \notin n)$ means $\neg m \in n$.
$\{o, p, q\}$ stands for the set whose only members are o, p, and q.

We can think of these as abbreviations. For example, we can now write $a \cap b \subseteq c$ instead of $\forall x((x \in a \wedge x \in b) \to x \in c)$.

One more short cut: for statements φ, ψ, χ we will write $(\varphi \wedge \psi \wedge \chi)$ instead of $((\varphi \wedge \psi) \wedge \chi)$ or $(\varphi \wedge (\psi \wedge \chi))$; similarly for $\vee$.

PROJECT #2. For the purposes of this exercise, assume that the only sets which exist in the entire universe are these: $a = \{\ \}$, $b = \{d, a\}$, $c = \{a, e\}$, $d = \{a, e, c\}$, $e = \{a\}$, and $f = \{a, b, c, d, e\}$. The following statements say something about the set x. In each case, there is exactly one of the sets a, b, c, d, e, f, which could be x. Find that set.

(i) $x \in e$ (ii) $x \notin f$ (iii) $d \subsetneq x$ (iv) $f \subseteq x$ (v) $x \subseteq a$
(vi) $\forall y(y \notin x)$ (vii) $\exists z(z \in x \wedge \forall w(w \in x \to w = z))$
(viii) $(x \in b \wedge x \notin d)$ (ix) $\neg(x \in c \to x \in b)$ (x) $\forall q(x \notin q)$
(xi) $e = \{a, x\}$ (xii) $x = \{\{\ \}\}$ (xiii) $(x \notin d \leftrightarrow x \subseteq d)$
(xiv) $\exists h \exists j(h \in j \wedge j \in d \wedge x \in h)$ (xv) $\forall n \neg(n \subsetneq x)$
(xvi) $\exists r(\forall u u \notin r \wedge r \notin x)$ (xvii) $\exists v(v \subsetneq x \wedge \forall k(k \subsetneq x \to k = v))$
(xviii) $\exists g \exists h \exists i \exists j(j \in i \wedge i \in h \wedge h \in g \wedge g \in x)$
(xix) $\exists y \exists z((y \in z \wedge z \in x) \wedge y \notin x)$
(xx) $\exists q \exists r \exists s((q \in x \wedge r \in x \wedge s \in x \wedge q \neq r \wedge r \neq s \wedge q \neq s)$
        $\wedge \ \forall t(t \in x \to (t = q \vee t = r \vee t = s)))$

We will occasionally use a richer language:

Definition. $\mathscr{L}^+$ is the language obtained from $\mathscr{L}$ by adding a constant symbol for each set.

Some Axioms of Zermelo–Fraenkel Set Theory (ZF)

Extension: Two sets are equal iff they have the same members.
Empty Set: There is a set with no elements.
Pair Set: If c and d are sets, then $\{c, d\}$ is a set.
Union: If d is a set of sets, then the union of these sets is a set.
Power Set: If d is a set, then the collection of all subsets of d is also a set (called the *power set* of d).
Regularity: If d is a set, then either $d = \{\ \}$ or else d has a member b such that $d \cap b = \{\ \}$.

1. Logic and Set Theory

The difficulty with earlier formulations of set theory was in allowing any definable collection to be a set. In this next axiom, we limit ourselves to definable collections which are already contained in a set (hence not too big).

Comprehension: If D is a set, then every definable part of D is also a set. If $\varphi(x)$ is a formula of $\mathscr{L}^+$, then the collection of all elements x of D such that $\varphi(x)$ is true is what we are calling a "definable part" of D. We write this as

$$\{x \in D \mid \varphi(x)\}.$$

The Axiom of Comprehension states that this is a set.

There are two additional axioms which we will add later.

PROJECT #3. Write the first six axioms of ZF in the language $\mathscr{L}$, using any abbreviations defined so far.

Important Note! We have introduced the languages $\mathscr{L}$ and $\mathscr{L}^+$ for use in certain specialized areas, the Axiom of Comprehension being the first example. We will continue to use English for all other purposes. While we could translate portions of our work into $\mathscr{L}$, there is really no reason to do so. It would be extremely difficult to read! Similarly, in writing proofs you are under no obligation to use $\mathscr{L}$. Proofs are meant to be understood and appreciated, so they should be written in whatever language is most appropriate.

PROJECT #4. Prove:

1.1. Theorem. *If A and B are sets, then $A \cap B$ is a set.*

1.2. Theorem. *If A and B are sets, then $A \cup B$ is a set.*

1.3. Theorem. *If A is a set, then $\{A\}$ is a set.*

1.4. Theorem. *If A is a set, then $A \notin A$.*

1.5. Theorem. *If A and B are sets and $A \in B$, then $B \notin A$.*

1.6. Theorem. *There is only one set with no members.*

1.7. Theorem. *The sets described in the Union axiom and the Power Set axiom are unique.*

Notation (Add to $\mathscr{L}$). Let $\emptyset$ stand for $\{\ \}$, the unique set with no members. For any set d, let $\cup d$ stand for the set described in the Union axiom. For any set d, let $P(d)$ stand for the set described in the Power Set axiom.

PROJECT #5. Our problem now is to define, for sets a and b, a set to represent the ordered pair $\langle a, b \rangle$. It must be a set (everything must be a set) and it must have the property that $\langle a, b \rangle = \langle c, d \rangle$ iff both $a = c$ and $b = d$. The following is a list of possibilities. Only one works. Find it, prove that it works and is a set, and show why the others fail to work:

$$\{a, b\}; \quad \{a, \{b\}\}; \quad \{\{a\}, \{b\}\}; \quad \{\{a\}, \{a, b\}\}; \quad \{a, b, \{a, b\}\}.$$

Definition. $A \times B$ is the collection of all ordered pairs $\langle a, b \rangle$ where $a \in A$ and $b \in B$.

Prove:

1.8. Theorem. *If A and B are sets, then $A \times B$ is a set.*

Notation. Add $\langle a, b \rangle$ and $A \times B$ to $\mathscr{L}$ as abbreviations.

Definition. A *function* f from A to B is a subset of $A \times B$ such that
(a) if $x \in A$ then there is a y such that $\langle x, y \rangle \in f$.
(b) If $\langle x, y \rangle$ and $\langle x, z \rangle$ are in f, then $y = z$.

Notation. In $\mathscr{L}$ we will write $f(x) = y$ to mean $\langle x, y \rangle \in f$.

Definition. If f is a function from A to B, A is called the *domain* of f. The *range* of f is $\{y \in B \mid \exists x f(x) = y\}$. f is *one-to-one* iff whenever $f(x) = f(y)$ then $x = y$. f is *onto* if the range of f is all of B. If $D \subseteq A$, then $f \restriction D$ (*f restricted to D*) is $\{\langle x, y \rangle \in f \mid x \in D\}$.

If f is a one-to-one function from A to B, then f^{-1}, the *inverse function*, is the set $\{\langle y, x \rangle \in B \times A \mid \langle x, y \rangle \in f\}$.

Prove:

1.9. Theorem. *If f is a function then the range of f is a set.*

1.10. Theorem. *If f is a function and D is a set, then $f \restriction D$ is a set.*

1. Logic and Set Theory

1.11. Theorem. *If f is a one-to-one function from A to B, then f^{-1} is a one-to-one function from B to A.*

Definition. A *relation* R on a set A is a subset of $A \times A$.

Notation. In $\mathscr{L}$, write aRb for $\langle a, b \rangle \in R$.

Definition. A relation R on A is *reflexive* if xRx for all $x \in A$. R is *symmetric* if xRy implies yRx. R is *transitive* if whenever xRy and yRz then xRz. R is an *equivalence relation* iff R is reflexive, transitive, and symmetric.

Notation. If R is an equivalence relation on A, $a \in A$, we will write $[a]_R$ for $\{b \in A \mid aRb\}$ (called the *equivalence class of a*).

PROJECT #6. Prove:

1.12. Theorem. *If R is an equivalence relation on A and $a \in A$, then $[a]_R$ is a set.*

1.13. Theorem. *If R is an equivalence relation on A and $a, b \in A$, then either $[a]_R = [b]_R$ or $[a]_R \cap [b]_R = \varnothing$.*

1.14. Theorem. *If R is an equivalence relation on A, then the collection of equivalence classes on A is a set.*

CHAPTER 2
The Natural Numbers

It is in the contemplation of the infinite that man attains his greatest good.

Giordano Bruno

The object of this chapter is to define a set to represent the numbers 0, 1, 2, ... To be complete, we must also show how to add and multiply these numbers and prove all the usual laws: commutative, associative, etc. The most important idea contained in our construction is that of mathematical induction.

To begin, we require an axiom which gives us a set large enough to construct the natural numbers. In fact, the axioms we have so far are too weak to guarantee us an infinite set.

Definition. For any set x, let $S(x) = x \cup \{x\}$.

Notation. Add S to $\mathscr{L}$.

Infinity: There is a set X such that

(1) $\emptyset \in X$
(2) if $y \in X$ then $S(y) \in X$
(3) if $y \in X$ and $y \neq \phi$, then $y = S(z)$ for some $z \in X$

Any set satisfying these conditions must contain at least $\emptyset$, $S(\emptyset)$, $S(S(\emptyset))$, $S(S(S(\emptyset)))$, ..., intuitively an infinite set. We postpone until Chapter 7 a complete discussion of "finite" and "infinite."

PROJECT #7. Prove:

2.1. Theorem. *If x is a set then $S(x)$ is a set.*

2.2. Theorem. *The conditions* (1), (2), *and* (3) *can be true about only one set.*

Definition. Let $\mathbb{N}$ (the natural numbers) be the unique set satisfying (1), (2), and (3).

Definition.

$0 = \emptyset$
$1 = S(0)$
$2 = S(1)$
$3 = S(2)$
$4 = S(3)$

Write out the sets 0, 1, 2, 3 and 4 using only the symbols, "{", "}", and ",".

Prove:

2.3. Theorem. $\mathbb{N}$ *satisfies Peano's Axioms*:

(a) $0 \in \mathbb{N}$.
(b) *For all* $k \in \mathbb{N}$ $S(k) \in \mathbb{N}$.
(c) $S(k) \neq 0$ *for all* $k \in \mathbb{N}$.
(d) *For all* $j, k \in \mathbb{N}$ $S(j) = S(k)$ *iff* $j = k$.
(e) *If* $X \subseteq \mathbb{N}$, *and* $0 \in X$, *and if for all* $k \in \mathbb{N}$ $k \in X$ *implies* $S(k) \in X$, *then* $X = \mathbb{N}$.

2.4. Theorem (Principle of Induction). *If a statement $\varphi(k)$ in $\mathscr{L}$ is*

(1) *true about 0, and*
(2) *whenever it is true about a number k, $k \in \mathbb{N}$, then it is also true about $S(k)$,*

then $\varphi(k)$ is true about all $k \in \mathbb{N}$.

2. The Natural Numbers

Defining Addition. We can describe addition as a function A from $\mathbb{N} \times \mathbb{N}$ to $\mathbb{N}$. Think of $2 + 3 = 5$, for example, as $A(\langle 2,3 \rangle) = 5$. At the very least, this function A should satisfy two properties:

(1) $A(\langle n, 0 \rangle) = n$, and
(2) $A(\langle n, S(k) \rangle) = S(A(\langle n, k \rangle))$

Think of $S(k)$ as $k + 1$. These laws say $n + 0 = n$ and $n + (k+1) = (n+k) + 1$.

Proving the existence of such a function is surprisingly tricky.

2.5. Theorem. *There is a function satisfying* (1) *and* (2).

The proof is given at the end of this chapter.

Notation. We will write $a +_\mathbb{N} b = c$ instead of $A(\langle a,b \rangle) = c$ (or worse, $\langle \langle a,b \rangle, c \rangle \in A$).

PROJECT #8. Prove:

2.6. Theorem. $+_\mathbb{N}$ *is associative.*

2.7. Theorem. $S(n) = n +_\mathbb{N} 1$ *for all* $n \in \mathbb{N}$.

2.8. Theorem. $0 +_\mathbb{N} n = n$ *for all* $n \in \mathbb{N}$.

PROJECT #9. Prove:

2.9. Theorem. $+_\mathbb{N}$ *is commutative.*

2.10. Theorem. $2 +_\mathbb{N} 2 = 4$.

Defining Multiplication. A multiplication function M from $\mathbb{N} \times \mathbb{N}$ to $\mathbb{N}$ should satisfy:

(1) $M(\langle n, 0 \rangle) = 0$, and
(2) $M(\langle n, S(k) \rangle) = M(\langle n, k \rangle) +_\mathbb{N} n$.

2.11. Theorem. *There is a function satisfying* (1) *and* (2).

Notation. We will write $a \cdot_\mathbb{N} b = c$ for $M(\langle a,b \rangle) = c$.

PROJECT #10. Prove:

2.12. Theorem. $n \cdot_\mathbb{N} (m +_\mathbb{N} p) = (n \cdot_\mathbb{N} m) +_\mathbb{N} (n \cdot_\mathbb{N} p)$ *for all* n, m, $p \in \mathbb{N}$.

2.13. Theorem. $n \cdot_\mathbb{N} 1 = n$ *for all* $n \in \mathbb{N}$.

2.14. Theorem. $\cdot_\mathbb{N}$ *is associative.*

PROJECT #11. Prove.

2.15. Theorem. $\cdot_\mathbb{N}$ *is commutative.*

2.16. Theorem. $(n +_\mathbb{N} m) \cdot_\mathbb{N} p = (n \cdot_\mathbb{N} p) +_\mathbb{N} (m \cdot_\mathbb{N} p)$ *for all* n, m, $p \in \mathbb{N}$.

2.17. Theorem. $2 \cdot_\mathbb{N} 2 = 4$.

Definition. A relation R on A is *irreflexive* iff $\neg xRx$ for all $x \in A$. R satisfies trichotomy iff whenever $x, y \in A$, then one and only one of the following is true: xRy, yRx, or $x = y$. R is called a *partial ordering* if it is irreflexive and transitive. A partial ordering is called a *linear ordering* if it satisfies trichotomy.

Definition. For all $x, y \in \mathbb{N}$, we will say $x <_\mathbb{N} y$ iff $x +_\mathbb{N} S(k) = y$ for some $k \in \mathbb{N}$.

PROJECT #12. Prove:

2.18. Theorem. *If* $x <_\mathbb{N} y$ *then* $x \in y$, *for all* $x, y \in \mathbb{N}$.

2.19. Theorem. $<_\mathbb{N}$ *defines a linear ordering on* $\mathbb{N}$.

PROOF OF THEOREM 2.5

Definition. If $m \in \mathbb{N}$ and A is a function satisfying:

(1) $A(\langle n, 0 \rangle) = n$, and
(2) $A(\langle n, S(k) \rangle) = S(A(\langle n, k \rangle))$ for all $k \in m$ and $n \in \mathbb{N}$, we will say that A is *good for adding numbers up to m*.

2. The Natural Numbers

We will prove first that for every m, there is a function good for adding numbers up to m. We do this by induction. First, for $m = 0$.

A function from $\mathbb{N} \times \mathbb{N}$ to $\mathbb{N}$ is a subset of $(\mathbb{N} \times \mathbb{N}) \times \mathbb{N}$. If we take just the collection of sets $\langle \langle n, 0 \rangle, n \rangle$ for all $n \in \mathbb{N}$ (this is a definable subset of $(\mathbb{N} \times \mathbb{N}) \times \mathbb{N}$), then this will satisfy (1) easily. Since there are no $k \in 0$, we don't have to worry about (2), hence this function is good for adding numbers up to 0.

Now suppose we have a function A good for adding numbers up to m. Let A^+ be the function consisting of:

(i) all ordered pairs in A, plus
(ii) the ordered pairs $\langle \langle n, S(m) \rangle, S(A(\langle n, m \rangle)) \rangle$, $n \in \mathbb{N}$.

This guarantees that A^+ is good for adding numbers up to $S(m)$, since if $k \in S(m)$, then either $k \in m$, and (i) shows that A^+ works as A does, or $k = m$, and (ii) insures that A^+ satisfies (2).

Finally, let $A^\#$ consist of all sets $\langle \langle n, k \rangle, d \rangle$ such that $A(\langle n, k \rangle) = d$ for some A which is good for adding numbers up to k.

Clearly $A^\#$ satisfies both (1) and (2). The domain of $A^\#$ is all of $\mathbb{N} \times \mathbb{N}$. The only problem is: is $A^\#$ actually a function? What if A_1 and A_2 are both good for adding numbers up to k (or more) and $A_1(\langle n, k \rangle) \neq A_2(\langle n, k \rangle)$ for some n? If this happens, we would have two different values for $A^\#(\langle n, k \rangle)$.

Let X be the set of such k. We will see that $X = \emptyset$, showing that $A^\#$ is indeed a function. First, $0 \in \mathbb{N} \setminus X$ since

$$A_1(\langle n, 0 \rangle) = n = A_2(\langle n, 0 \rangle).$$

Further, if $k \in \mathbb{N} \setminus X$, then $S(k) \in \mathbb{N} \setminus X$, since

$$A_1(\langle n, S(k) \rangle) = S(A_1(\langle n, k \rangle)) = S(A_2(\langle n, k \rangle)) = A_2(\langle n, S(k) \rangle).$$

By Theorem 2.3(e), $\mathbb{N} \setminus X = \mathbb{N}$, so $X = \emptyset$. □

A very similar argument proves Theorem 2.11.

CHAPTER 3
The Integers

God made the integers. All else is the work of man.

Leopold Kronecker

In this chapter we will construct a set to represent the positive and negative integers. As before, we will define addition and multiplication. In addition to the properties proved for $\mathbb{N}$, we will now have additive inverses. The key idea in our construction is the use of equivalence classes.

We begin by defining a relation $\sim$ on $\mathbb{N} \times \mathbb{N}$:

Definition. $\langle a, b \rangle \sim \langle c, d \rangle$ iff $a +_\mathbb{N} d = b +_\mathbb{N} c$.

PROJECT #13. Prove:

3.1. Theorem. $\sim$ *defines an equivalence relation on* $\mathbb{N} \times \mathbb{N}$.

Definition. $\mathbb{Z}$ (the integers) = the collection of all equivalence classes of $\sim$.

Prove:

3.2. Theorem. $\mathbb{Z}$ *is a set.*

Invent: $+_\mathbb{Z}$, addition for $\mathbb{Z}$.

PROJECT #14. Prove:

3.3. Theorem. $+_\mathbb{Z}$ *is commutative and associative.*

Define $0_\mathbb{Z} \in \mathbb{Z}$ and prove:

3.4. Theorem. $0_\mathbb{Z} +_\mathbb{Z} a = a$ *for all* $a \in \mathbb{Z}$.

For all $a \in \mathbb{Z}$ define $^-(a)_\mathbb{Z}$, and prove:

3.5. Theorem. $a +_\mathbb{Z} {}^-(a)_\mathbb{Z} = 0_\mathbb{Z}$ *for all* $a \in \mathbb{Z}$.

Define $<_\mathbb{N}$ on $\mathbb{Z}$ and prove:

3.6. Theorem. $<_\mathbb{N}$ *is a linear ordering on* $\mathbb{Z}$.

PROJECT #15. Define $\cdot_\mathbb{Z}$ on $\mathbb{Z}$, and prove:

3.7. Theorem. $\cdot_\mathbb{N}$ *is commutative, associative, and distributive over* $+_\mathbb{Z}$.

Define $1_\mathbb{Z} \in \mathbb{Z}$, and prove:

3.8. Theorem. $1_\mathbb{Z}$ *is the identity for* $\cdot_\mathbb{Z}$.

CHAPTER 4
The Rationals

Numbers are intellectual witnesses that belong only to mankind, and by whose means we can achieve an understanding of words.

Honoré de Balzac

Our next goal is to construct the rational numbers. The method is very much like that of the previous chapter.

PROJECT #16. Define $\mathbb{Q}$ (the rational numbers).

PROJECT #17. Define $+_\mathbb{Q}$ and prove:

4.1. Theorem. $+_\mathbb{Q}$ *is commutative and associative.*

Define $0_\mathbb{Q}$. Define for all $x \in \mathbb{Q}$ $^-(x)_\mathbb{Q}$ and prove:

4.2. Theorem. $0_\mathbb{Q}$ *is the identity for* $+_\mathbb{Q}$, *and for each* $x \in \mathbb{Q}$, $^-(x)_\mathbb{Q}$ *is its additive inverse.*

PROJECT #18. Define $\cdot_\mathbb{Q}$ and prove:

4.3. Theorem. $\cdot_\mathbb{Q}$ *is commutative, associative, and distributive over* $+_\mathbb{Q}$.

Define $1_\mathbb{Q}$. Define for all $x \in \mathbb{Q}$, $x \neq 0_\mathbb{Q}$ $(1/x)_\mathbb{Q}$ and prove:

4.4. Theorem. $1_\mathbb{Q}$ *is the identity for* $\cdot_\mathbb{Q}$, *and for each* $x \in \mathbb{Q}$, $x \neq 0_\mathbb{Q}$, $(1/x)_\mathbb{Q}$ *is its multiplicative inverse.*

PROJECT #19.

Definition. $[\langle a, b \rangle]_\approx$ is *positive* iff $0_\mathbb{Z} <_\mathbb{Z} a \cdot_\mathbb{Z} b$, and *negative* iff $a \cdot_\mathbb{Z} b <_\mathbb{Z} 0_\mathbb{Z}$.

Prove that "positive" and "negative" are well-defined.

Definition. For $r, s \in \mathbb{Q}$, $r <_\mathbb{Q} s$ iff $s +_\mathbb{Q} {}^-(r)_\mathbb{Q}$ is positive.

4.5. Theorem. $<_\mathbb{Q}$ *is a linear ordering.*

CHAPTER 5
The Real Numbers

> Young man, in mathematics you don't understand things, you just get used to them.
>
> John von Neumann

We complete our construction of the standard number systems with Dedekind's approach to the real numbers. For various reasons, there is a lot more work involved in this task, so we will limit ourselves to the definition of $\mathbb{R}$, $+_\mathbb{R}$, and $0_\mathbb{R}$, and some examination of the difficulties of proceeding further.

Definition. A *schnitt* is a subset $r \subseteq \mathbb{Q}$ such that:

(1) $q \in r$ and $p <_\mathbb{Q} q$ imply $p \in r$;
(2) r has no greatest element;
(3) $r \neq \emptyset$;
(4) $r \neq \mathbb{Q}$.

Definition. $\mathbb{R}$ is the collection of all schnitts.

PROJECT #20. Prove:

5.1. Theorem. $\mathbb{R}$ *is a set.*

Define $<_\mathbb{R}$ and prove:

5.2. Theorem. $<_\mathbb{R}$ *is a linear ordering.*

Definition. $u \in \mathbb{R}$ is an *upper bound* for $X \subseteq \mathbb{R}$ iff $b \leqslant_\mathbb{R} u$ for all $b \in X$. An upper bound u is the *least upper bound* for X iff whenever c is an upper bound for X, $u \leqslant_\mathbb{R} c$.

Prove:

5.3. Theorem (Continuity of the Reals). *If $X \subseteq \mathbb{R}$, $X \neq \varnothing$, has an upper bound, then it has a least upper bound.*

Definition. For $r, s \in \mathbb{R}$, $r +_\mathbb{R} s = \{z \in \mathbb{Q} \mid z$ is the sum of a member of r and a member of $s\}$.

PROJECT #21. Prove:

5.4. Theorem. *For $r, s \in \mathbb{R}$, $r +_\mathbb{R} s$ is a schnitt.*

PROJECT #22. Prove:

5.5. Theorem. $+_\mathbb{R}$ *is commutative and associative.*

Define $0_\mathbb{R}$ and prove:

5.6. Theorem. $0_\mathbb{R}$ *is the identity for* $+_\mathbb{R}$.

PROJECT #23 (No Proofs).
 Define $^-(r)_\mathbb{R}$ for all $r \in \mathbb{R}$.
 Define $\cdot_\mathbb{R}$.
 Define $(1/r)_\mathbb{R}$ for all $r \in \mathbb{R}$, $r \neq 0_\mathbb{R}$.

CHAPTER 6
The Ordinals

> A mathematician is a blind man in a dark room looking for a black hat which isn't there.
>
> Charles Darwin

We wish to extend $\mathbb{N}$, our set of counting numbers, to a larger class of numbers we can use to count infinite sets. These will be our first type of infinite number, and they will be used to measure the "lengths" of large sets.

Definition. A linear ordering on X is a *well-ordering* if every non-empty subset of X has a least element. A set with a well-ordering is said to be *well-ordered*.

Definition. An *ordinal* X is a set with the properties:

(1) $\in$ is a well-ordering on X, and
(2) if $a \in b$ and $b \in X$ then $a \in X$.

It is customary to use Greek letters, $\alpha, \beta, \gamma, \ldots$ to represent ordinals.

PROJECT #24. Prove:

6.1. Theorem. $\emptyset$ *is an ordinal.*

6.2. Theorem. $\mathbb{N}$ *is an ordinal.*

6.3. Theorem. *If α is an ordinal, then $S(\alpha)$ is an ordinal.*

6.4. Theorem. *If α is an ordinal and $b \in \alpha$, then b is an ordinal.*

Definition. It is customary in set theory to write ω for $\mathbb{N}$.

PROJECT #25. Prove:

6.5. Theorem. *For α, β, ordinals, $\alpha \subsetneq \beta \leftrightarrow \alpha \in \beta$.*

6.6. Theorem. $\in$ *is a linear ordering on the ordinals.*

6.7. Theorem. $\in$ *is a well-ordering on the ordinals.*

6.8. Theorem. *If A is a set of ordinals, then $\cup A$ is an ordinal and is the least upper bound of A.*

6.9. Theorem. *There is no largest ordinal.*

6.10. Theorem. *The collection of all ordinals is not a set.*

Definition. If an ordinal $\alpha = S(\beta)$ for some β, then α is a *successor* ordinal. If $\alpha \neq 0$ is not a successor ordinal, it is a *limit* ordinal.

PROJECT #26. Prove:

6.11. Theorem. ω *is the least limit ordinal.*

6.12. Theorem (Transfinite Induction). *Given a formula φ of $\mathscr{L}^+$ with the property that for all ordinals β: if for every $\alpha \in \beta$ $\varphi(\alpha)$ is true, then $\varphi(\beta)$ is also true, then $\varphi(\beta)$ is true for every β.*

We need now (for Theorem 6.13) the last axiom of Zermelo–Frankel set theory.

Replacement: Every definable mapping whose domain is a set is a function.
 Specifically, if $\varphi(x, y)$ is a formula in $\mathscr{L}^+$, then φ "defines" a mapping if for all x there is a unique y such that $\varphi(x, y)$ is true. The

6. The Ordinals

Axiom of Replacement states that if D is a set, then the restriction of this mapping to D is a *function*, that is, $\{\langle x,y\rangle | x\in D \text{ and } \varphi(x,y))\}$ is a set.

One benefit of this axiom is that we don't have to distinguish between mappings defined by formulas (which don't have to be sets) and functions (which must be sets). If a formula $\varphi(x,y)$ defines a mapping (that is, for all x there is a unique y such that $\varphi(x,y)$ is true) then we can represent it as a function f with the understanding that this makes sense only if we restrict the domain of f to a set. The successor mapping, $S(x) = x \cup \{x\}$ is a good example. It is not really a function, for its domain includes all sets. Since it is definable, however, it is a function whenever its domain is restricted to a set (for example, $\mathbb{N}$).

This axiom completes ZF. All our theorems have been proved using only ZF. Later we will introduce a few axioms that have been considered important, and occasionally, worth adding to ZF. Not meeting with universal acceptance, they have not been added, but they continue to draw interest from set theorists, algebraists, analysts, topologists, philosophers, and mathematicians in general.

Definition. A collection of statements is *consistent* iff no contradiction can be deduced from the statements.

Is ZF consistent? We don't know and probably never will! A famous theorem proved by Kurt Gödel in 1931 (the second Incompleteness Theorem) states that any consistent collection of statements which has certain common characteristics cannot prove its own consistency. Of course, if ZF is *not* consistent, then it can prove *anything* (for the same reason that a false statement implies anything—see Chapter 1); in particular, it could then prove that it *is* consistent!

We will assume for the rest of this outline that ZF is consistent.

This is a reasonable assumption for most of us. For most platonists, ZF consists of statements that are intuitively true about sets. It is correct. For formalists, ZF is either consistent or it is not. If it is not, then everything follows from it. We need only provide proofs if ZF is consistent. Finally, for the practical researcher, no one has found an inconsistency in decades of investigation, so it seems fairly safe!

Definition. A well-ordered set $\langle B, <_B \rangle$ has *order-type* α, α an ordinal, iff there is a one-to-one function f from α onto B which is *order-preserving* that is, $\beta \in \gamma$ implies $f(\beta) <_B f(\gamma)$ for all $\beta, \gamma \in \alpha$.

6.13. Theorem. *Every well-ordered set has a unique order-type.*

The proof is given at the end of this chapter.
Now we can construct the arithmetic of the ordinals:

Definition. For ordinals α, β, $\alpha +_o \beta$ is the order-type of the ordering:

$$(\text{―――――}\overset{\alpha}{\text{―――――}}\text{―――――})(\text{―――――}\overset{\beta}{\text{―――――}}\text{―――――})$$

[In detail, let $<_+$ be the ordering on $\alpha \cup (1 \times \beta)$ defined by: $a <_+ b$ iff either:

(1) $a, b \in \alpha$ and $a \in b$
(2) $a \in \alpha, b \in (1 \times \beta)$, or
(3) $a, b \in (1 \times \beta)$, $a = \langle 0, c \rangle$, $b = \langle 0, d \rangle$, and $c \in d$.]

Definition. For ordinals α, β, $\alpha \cdot_o \beta$ is the order-type of the ordering:

$$\left[(\text{――}\overset{\alpha}{\text{――}}\text{――})(\text{――}\overset{\alpha}{\text{――}}\text{――})\underset{\beta}{(\text{――}\overset{\alpha}{\text{――}}\text{――})}\text{――――} \right]$$

[In detail: let $<_o$ be the ordering on $\alpha \times \beta$ defined by: $\langle a, b \rangle <_o \langle c, d \rangle$ iff either:

(1) $b \in d$, or
(2) $b = d$ and $a \in c$.]

PROJECT #27 (No Proofs). Determine whether or not $+_o$ has the properties of commutativity, associativity. Does it have an identity? Is the right cancellation law true? (If $\alpha +_o \beta = \gamma +_o \beta$, must $\alpha = \gamma$?) If $+_o$ is not commutative, we also ask if the left cancellation law for $+_o$ is true.

Determine whether or not $\cdot_o$ has the properties of commutativity or associativity. Does it have an identity? Do the cancellation laws hold? Is $\cdot_o$ distributive over $+_o$?

Notation. It is standard practice to write $\alpha < \beta$ for $\alpha \in \beta$, since $\in$ is our well-ordering on the ordinals.

6. The Ordinals

THE PROOF OF THEOREM 6.13. We need first an important theorem:

6.14. Theorem on Recursive Definitions. *If f is any function, then there is a function g such that for all ordinals α:*

$$g(\alpha) = a \quad \text{iff} \quad a = f(\{g(\beta) | \beta < \alpha\}).$$

This is a theorem about building functions level-by-level. The functions $+_\mathbb{N}$ and $\cdot_\mathbb{N}$ were examples of this. Another example follows in the proof of 6.13. Others will come in Chapters 8 and 9.

PROOF. We can define g by a formula $\varphi(x, y)$ in $\mathscr{L}^+$:

"x is an ordinal and there is a function h satisfying:

$$h(\gamma) = f(\{h(\beta) | \beta < \gamma\}), \quad \text{for all } \gamma \leqslant x \qquad (*)$$

and $h(x) = y$."

This defines a mapping, for suppose for some ordinal α and sets y, z, both $\varphi(\alpha, y)$ and $\varphi(\alpha, z)$ are true, and $y \neq z$. Then there are two functions h_1 and h_2 satisfying $(*)$ with $h_1(\alpha) = y$ and $h_2(\alpha) = z$. Let α be the least such ordinal. Then

$$x = h_1(\alpha) = f(\{h_1(\beta) | \beta < \alpha\})$$
$$= f(\{h_2(\beta) | \beta < \alpha\})$$
$$= h_2(\alpha) = y$$

—a contradiction. By Replacement, φ defines a function g, and g satisfies the requirements of the theorem.

Now to prove 6.13. Suppose B is well-ordered by $<_B$. We apply the theorem on recursive definitions to f where $f(x)$ is defined to be the $<_B$-least element not in x (if there is one—it doesn't matter what $f(x)$ is otherwise). Let g be the mapping from 6.14 satisfying:

$$g(\alpha) = a \quad \text{iff} \quad a = f(\{g(\beta) | \beta < \alpha\}).$$

This implies that g is both one-to-one and order-preserving. Since it is one-to-one, the inverse function, g^{-1} defined by:

$$g^{-1}(a) = \alpha \quad \text{iff} \quad g(\alpha) = a$$

is also a well-defined mapping. Since the domain of g^{-1} is contained in B, it is a set by Comprehension. It follows from Replacement that g^{-1} is a function, and hence g as well. The domain of g, a set of ordinals, must be an ordinal, for if $g(\alpha)$ is defined and $\beta < \alpha$, then $g(\beta)$ is defined also. Let δ be the domain of g and let $S \subseteq B$ be the range. It only remains to show that $S = B$. But if $B \setminus S$ is nonempty, then $f(S) \in B$ and so

$$g(\delta) = f(\{g(\beta) | \beta < \delta\}) = f(S)$$

and so δ is in the domain of g, i.e., $\delta \in \delta$. This is, of course, impossible. □

CHAPTER 7
The Cardinals

My bounty is as boundless as the sea,
My love as deep; the more I give to thee
The more I have, for both are infinite.

William Shakespeare

We develop in this chapter a second set of infinite numbers to measure the *size* (as opposed to the *length*) of infinite sets.

Definition. For sets A and B, we say:

(1) $\|A\| = \|B\|$ iff there is a function from A to B which is one-to-one and onto,

(2) $\|A\| \leq \|B\|$ iff there is a function from A to B which is one-to-one, and

(3) $\|A\| < \|B\|$ iff $\|A\| \leq \|B\|$ but not $\|B\| \leq \|A\|$.

PROJECT #28. Prove:

7.1. Theorem. $\|\omega\| \leq \|S(\omega)\| \leq \|\omega\|$.

7.2. Theorem. $\|\omega\| \leq \|\mathbb{Z}\| \leq \|\omega\|$.

7.3. Theorem. $\|\omega\| \leq \|\mathbb{Q}\| \leq \|\omega\|$.

PROJECT #29. Prove:

7.4. Theorem. (The Shroeder–Bernstein Theorem). *If* $\|A\| \leq \|B\| \leq \|A\|$ *then* $\|A\| = \|B\|$.

7.5. Theorem. $\|\omega\| = \|\mathbb{Z}\| = \|\mathbb{Q}\|$.

PROJECT #30. Prove:

7.6. Theorem (Georg Cantor, 1874). $\|A\| < \|P(A)\|$ *for all sets* A.

7.7. Theorem (Georg Cantor, 1874). $\|\omega\| < \|\mathbb{R}\|$.

Definition. An ordinal α is called a *cardinal* if $\|\beta\| < \|\alpha\|$ for all $\beta < \alpha$. It is customary to use the Hebrew letter $\aleph$ to represent cardinals.

7.8. Theorem. *If A is a set of cardinals, then $\cup A$ is a cardinal.*

7.9. Theorem. *$S(\omega)$ is not a cardinal.*

7.10. Theorem. *The following statements are equivalent*:

(1) The Axiom of Choice (AC): *For any set A, there is a function f (called a* choice function*) on A such that $f(x) \in x$ for all $x \in A$, $x \neq \emptyset$.*
(2) Zorn's Lemma: *If $P \neq \emptyset$ is a set partially ordered by $<_p$ such that every* chain (*that is, a set $C \subseteq P$ such that $<_p$ is a linear order on C*) *is* bounded (*that is, there is some $x \in P$ such that for all $y \in C$, $y \leq_p x$*), *then P has a maximal element* (*that is, an element $x \in P$ such that for no y in P is $x <_p y$*).
(3) Zermelo's Theorem, *or* The Well-ordering Theorem: *Every set can be well-ordered.*

We will prove this theorem in Chapter 9.

The Axiom of Choice is considered by some mathematicians to be an essential part of set theory and mathematics. Others regard it variously as optional, irrelevant, or actually false. When added to ZF, the resulting system is called *ZFC*.

Early in the history of set theory, there were attempts to prove (or disprove) AC from ZF. These attempts failed in a very spectacular way.

7. The Cardinals

7.11. Theorem (Kurt Gödel, 1936). *It is impossible to disprove* AC.

7.12. Theorem (Paul Cohen, 1963). *It is impossible to prove* AC.

These two theorems are startling, to say the least. Mathematicians had been prepared for something of the sort by Gödel's first Incompleteness Theorem (1931) in which he showed that for any consistent set of statements which had certain common characteristics there were statements that could neither be proved nor disproved. This, however, was the first really concrete example.

The theorems above are very deep, but in the next chapter we will get an idea of how they are proved.

PROJECT #31.

Definition. A set X is *finite* iff for some $n \in \mathbb{N}$, $\|n\| = \|X\|$. X is *infinite* iff it is not finite.

A set X is *Dedekind infinite* iff there is a one-to-one function from X onto a proper subset of X. X is *Dedekind finite* iff it is not Dedekind infinite.

Prove:

7.13. Theorem. *Each* $n \in \mathbb{N}$ *is Dedekind finite.*

7.14. Theorem. *If X is finite then it is Dedekind finite.*

Using ZFC prove:

7.15. Theorem. *If X is infinite then there is a function mapping ω one-to-one into X.*

7.16. Theorem. *If X is infinite then it is Dedekind infinite.*

Definition. A set X is *countable* iff $\|X\| \leq \|\omega\|$.

PROJECT #32. Using ZFC prove:

7.17. Theorem. *The countable union of countable sets is countable.*

7.18. Theorem. *If A is countable, then $A \times A$ is countable.*

7.19. Theorem. *There is no largest cardinal.*

PROJECT #33. Using only ZF prove:

7.20. Theorem. *If f is a function which maps a set X one-to-one onto an ordinal β, then there is a well-ordering of X of order-type β.*

7.21. Theorem (Hartog's Theorem). *There is no largest cardinal.*

Definition. For any set A, $\|A\|$ is the unique cardinal number $\aleph$ such that $\|A\| = \|\aleph\|$ (if there is one). For $\aleph$ a cardinal, $\|\aleph\| = \aleph$. We will denote the next largest cardinal after $\aleph$ by $\aleph^+$.

Definition.

$$\aleph_0 = \omega$$
$$\aleph_1 = (\aleph_0)^+$$
$$\aleph_{n+1} = (\aleph_n)^+, \quad \text{for all } n \in \omega$$
$$\aleph_\omega = \cup\{\aleph_n | n \in \omega\}$$
$$\aleph_{\omega+1} = (\aleph_\omega)^+$$

and in general,

$$\aleph_\lambda = \text{the } \lambda^{\text{th}} \text{ infinite cardinal.}$$

Finally, $\mathfrak{c} = \|P(\omega)\| = \|\mathbb{R}\|$ ($\mathfrak{c}$ is a German "c" for "continuum").

The Continuum Hypothesis. (CH). $\mathfrak{c} = \aleph_1$

The Generalized Continuum Hypothesis. (GCH). $\|P(\aleph)\| = \aleph^+$, for all cardinals $\aleph$, $\aleph_0 \leqslant \aleph$.

It was Cantor who formulated CH and GCH. He thought it likely CH was true and tried to prove it. It is a very natural idea. By Theorem 7.7, $\|\mathbb{R}\|$ is larger than $\omega = \aleph_0$, how much larger?

7.22. Theorem (Kurt Gödel, 1936). *It is impossible to disprove either CH or GCH.*

7.23. Theorem (Paul Cohen, 1963). *It is impossible to prove either CH or GCH.*

CHAPTER 8
The Universe

> —listen:there's a hell
> of a good universe next door;let's go
>
> E. E. Cummings

We now explore some pure set theory, examining the structure of the universe of sets. A crucial concept will be that of a set which in itself is a universe of sets, that is, all the axioms of ZF are true about the members of this set.

Definition.

(a) $V(0) = \emptyset$
(b) $V(\alpha + 1) = P(V(\alpha))$ for all ordinals α
(c) $V(\lambda) = \bigcup \{V(\alpha) | \alpha < \lambda\}$ for all limit ordinals λ.

8.1. Theorem. $V(\alpha)$ *is defined and unique for all ordinals* α.

The proof follows exactly the proof of 6.14, the Theorem on Inductive Definitions, but define the formula φ by:

"x is an ordinal and there is a function h satisfying:

$h(0) = 0$,

$h(\alpha +_o 1) = P(h(\alpha))$ for all $\alpha < x$,

$h(\lambda) = \cup\{h(\beta)|\beta < \lambda\}$ for all limit ordinals $\lambda \leqslant x$,

and

$h(x) = y$."

Note that as a consequence, not only is V a function, but we can express "$z \in V(\alpha)$" in $\mathscr{L}$.

It is customary to write V_α for $V(\alpha)$.

PROJECT #34. Write out:

$$V_0, V_1, V_2, \text{ and } V_3$$

using only the symbols "{", "}", and ", ".

Prove:

8.2. Theorem. *For all α, $x \in y \in V_\alpha \to x \in V_\alpha$.*

8.3. Theorem. *$\delta < \alpha \to V_\delta \subseteq V_\alpha$, for all ordinals α and δ.*

PROJECT #35. Prove:

8.4. Theorem. *Every set is in some V_α.*

Definition. A cardinal κ is *regular* if whenever $X \subseteq \kappa$, $\|X\| < \kappa$, then $\cup X < \kappa$. If κ is not regular we say it is *singular*.

PROJECT #36. Prove:

8.5. Theorem. *ω is regular.*

8.6. Theorem. *$\aleph_\omega$ is singular.*

8.7. Theorem. *If we assume AC, then $\aleph_1$ is regular.*

Definition. A cardinal $\kappa > \omega$ is *strongly inaccessible* if it is regular and if whenever $\|X\| < \kappa$, $\|P(X)\| < \kappa$. Note that $\|P(X)\|$ makes sense only if we can well-order $P(X)$.

8. The Universe

PROJECT #37. Prove:

8.8. Theorem. ω, $\aleph_1$, $\aleph_6$ and $\aleph_\omega$ *are not strongly inaccessible.*

8.9. Theorem (AC). *If κ is strongly inaccessible and $\alpha < \kappa$ then $\|V_\alpha\| < \kappa$.*

PROJECT #38. Prove using ZFC:

8.10. Theorem. *If κ is strongly inaccessible, then all the axioms of ZF are true in V_κ.*

Definition. If T is a set of axioms (called a *theory*) and every axiom of T is true in a set X, then X is a *model* of T.

Theorem 8.10 states that if κ is strongly inaccessible then V_κ is a model for ZF.

Definition. If T is a set of statements, Con(T) is the assertion that T is consistent.

EXAMPLES:

(1) We have been assuming (starting in Chapter 6) that Con(ZF) is true.
(2) Gödel's second Incompleteness Theorem (Chapter 6) stated that if Con(T), then Con(T) can't be proved from T (for suitable T).
(3) An earlier theorem of Gödel connects the ideas of models and consistency. The Completeness Theorem (1930) states that if Con(T) is true then T has a model (the converse is trivially true).

We introduce another axiom:

SI. There exists a strongly inaccessible cardinal.

(4) In the light of (3), Theorem 8.10 says that ZFC + SI implies Con(ZF). In fact, with more work we could prove ZF + SI implies Con(ZF).
(5) Theorems 7.11 and 7.12 can be rewritten as:

$$\text{Con(ZF)} \to \text{Con(ZF + AC)}$$

$$\text{Con(ZF)} \to \text{Con(ZF + } \neg\text{AC)}.$$

PROOF. Assuming that ZF is consistent, then ZF cannot disprove AC or $\neg$AC, and so both ZF$\cup${AC} and ZF$\cup${$\neg$AC} are consistent axiom systems.)

These are called *relative* consistency proofs. We can't prove Con(ZFC), but we can prove that it follows from Con(ZF). Early set theorists worried that adding AC to ZF might create an inconsistency. Gödel's theorem says not to worry, that if the combination of ZF and AC is inconsistent, then ZF is inconsistent all by itself. This means that AC is a relatively "safe" axiom. So is $\neg$AC, via Cohen's theorem.

How does one prove relative consistency? Gödel's theorem, that Con(ZF) implies Con(ZF + CH) proceeds as follows: if ZF is consistent, then there is a model M for it. Gödel then found that any such model has a (possibly smaller) model, L, inside it which satisfies both ZF and CH. Since it has a model, ZF + CH must be consistent. Cohen proved his theorem by taking Gödel's model L and delicately expanding it to form a model N in which ZF and $\neg$CH are both true, hence Con(ZF + $\neg$CH).

What about the axiom SI? How safe is it?

8.11. Theorem. Con(ZF) *does* not *imply* Con(ZF + SI). *In other words, we* can't *prove that we* can't *disprove* SI!

PROOF. Consider $T =$ ZF + Con(ZF) and suppose that Con(ZF) $\to$ Con(ZF + SI).

Then $T \to$ ZF + Con(ZF + SI)

$\to$ ZF + Con(ZF + Con(ZF)) (by (4) above)

$\to$ Con(ZF + Con(ZF))

$\to$ Con(T)

—but T can't imply its own consistency by the second Incompleteness Theorem. □

CHAPTER 9
Choice and Infinitesimals

I will not go so far as to say that to construct a history of thought without profound study of the mathematical ideas of successive epochs is like omitting Hamlet from the play which is named after him. That would be claiming too much. But it is certainly analogous to cutting out the part of Ophelia. The simile is singularly exact. For Ophelia is quite essential to the play, she is charming—and a little mad.

A. N. Whitehead

We prove here Theorem 7.10 which offers three equivalent forms of the Axiom of Choice. We then use AC to construct a system of numbers called the Hyperreal numbers ($\mathbb{HR}$). This system extends $\mathbb{R}$ as $\mathbb{R}$ extended $\mathbb{Q}$ and $\mathbb{Q}$ extended $\mathbb{Z}$. $\mathbb{HR}$ contains both infinite numbers and infinitesimals.

PROJECT #39. Prove:

9.1. Theorem. *The Axiom of Choice implies Zermelo's Theorem.*

9.2. Theorem. *Zermelo's Theorem implies Zorn's Lemma.*

PROJECT #40. Prove:

9.3. Theorem. *Zorn's Lemma implies the Axiom of Choice.*

This completes the proof of Theorem 7.10.

Definition. $\mathcal{U} \subseteq P(\omega)$ is a (non-principal) *ultrafilter on* ω iff
(a) if $A, B \in \mathcal{U}$ then $A \cap B \in \mathcal{U}$,
(b) if $\|A\| < \omega$ then $\omega \setminus A \in \mathcal{U}$, and
(c) for all $A \subseteq \omega$, either A or $\omega \setminus A$ is in $\mathcal{U}$.

Prove using ZFC:

9.4. Theorem. *There is an ultrafilter on ω.*

For the rest of this chapter, let $\mathcal{U}$ be an ultrafilter on ω. Assume, furthermore, all the usual facts about the real numbers which we did not prove in Chapter 5.

Definition. Let $\chi = \{f \mid f \text{ is a function from } \omega \text{ to } \mathbb{R}\}$, and define a relation, $\approx$, on χ by:

$$f \approx g \quad \text{iff} \quad \{n \in \omega \mid f(n) = g(n)\} \in \mathcal{U}$$

PROJECT #41. Prove:

9.5. Theorem. $\approx$ *is an equivalence relation.*

Definition. $\mathbb{HR}$ (the hyperreal numbers) is the collection of all equivalence classes of $\approx$.

Definition. $[f]_\approx <_{\mathsf{HR}} [g]_\approx$ iff $\{n \in \omega \mid f(n) <_{\mathbb{R}} g(n)\} \in \mathcal{U}$.

Prove:

9.6. Theorem. $<_{\mathsf{HR}}$ *is a well-defined linear ordering.*

Definition. For each $r \in \mathbb{R}$, let $f_r \in \chi$ be the function:

$$f_r(n) = r \quad \text{for all } n \text{ (a constant function)}.$$

Definition. $x \in \mathbb{HR}$ is *infinite* iff either
(1) $x <_{\mathsf{HR}} [f_r]_\approx$ for all $r \in \mathbb{R}$, *or*
(2) $x >_{\mathsf{HR}} [f_r]_\approx$ for all $r \in \mathbb{R}$.

9. Choice and Infinitesimals

Prove:

9.7. Theorem. $\mathbb{HR}$ *contains infinite numbers.*

Definition. $x \in \mathbb{HR}$ is an *infinitesimal* iff
(1) $x \neq [f_0]_\approx$, and
(2) $x <_{\mathsf{HR}} [f_r]_\approx$ for all $r \in \mathbb{R}$, $r >_\mathbb{R} 0_\mathbb{R}$, and
(3) $x >_{\mathsf{HR}} [f_r]_\approx$ for all $r \in \mathbb{R}$, $r <_\mathbb{R} 0_\mathbb{R}$.

Prove:

9.8. Theorem. $\mathbb{HR}$ *contains infinitesimals.*

The Hyperreals were developed in 1960 by Abraham Robinson as an alternative approach to the calculus. The early practitioners of the calculus (17th–19th centuries) spoke glibly of infinitely large and infinitely small quantities. While the ideas were extremely fruitful, there was considerable uneasiness that these notions seemed to have no foundation (mathematical or otherwise). Great minds would use infinitesimals one way and derive true theorems. Others would reason similarly and arrive at contradictions. Leibniz, one of the original discoverers of the calculus, always insisted that infinite quantities could be treated just like finite ones. There were limitations, however, and it was never clear what they were. Eventually these ideas were replaced by the theory of limits.

Robinson rescued the method of infinitesimals and in doing so discovered what the "limitations" were.

Definition. Let $\mathscr{L}_\mathbb{R}$ be the mathematical language constructed like $\mathscr{L}$, but with three changes:
(1) the symbol "∈" is not included,
(2) symbols for $+_\mathbb{R}$ and $\cdot_\mathbb{R}$ are included, and
(3) a constant symbol for each real number is included, as well as a function symbol for each real function.

Robinson proved:

9.9. Theorem. *If φ is a statement written in $\mathscr{L}_\mathbb{R}$, then φ is true about $\mathbb{R}$ iff and only if φ is true about $\mathbb{HR}$.*

The limitation, in other words, is that we must restrict ourselves to the language $\mathscr{L}_\mathbb{R}$.

PROJECT #42. (No Proofs). Define $+_{H\mathbb{R}}$ and $\cdot_{H\mathbb{R}}$.

Let $H, K \in H\mathbb{R}$ be positive and infinite. Let $I, J \in H\mathbb{R}$ be positive and infinitesimal. Determine whether the following statements are true or false:

(1) $H +_{H\mathbb{R}} K$ must be infinite.
(2) $I \cdot_{H\mathbb{R}} J$ must be infinitesimal.
(3) $I +_{H\mathbb{R}} J$ must be infinitesimal.
(4) $H \cdot_{H\mathbb{R}} [f_{.01}]_\approx$ must be infinite.
(5) $I \cdot_{H\mathbb{R}} [f_{100}]_\approx$ must be infinitesimal.
(6) There is a largest finite hyperreal.
(7) There is a largest infinite hyperreal.
(8) There is a smallest infinite, positive hyperreal.
(9) $\sin^2(H) +_{H\mathbb{R}} \cos^2(H) = [f_1]_\approx$.
(10) Every bounded subset of $H\mathbb{R}$ has a least upper bound.

The hyperreals $[f_r]_\approx$ constitute an exact copy of $\mathbb{R}$ inside $H\mathbb{R}$. In this sense, $H\mathbb{R}$ is an extension of $\mathbb{R}$. $H\mathbb{R}$ contains other numbers as well, infinite numbers and infinitesimals. The following theorem offers a clearer picture of the order structure of $H\mathbb{R}$.

9.10. Theorem. *Every finite hyperreal is infinitely close to a unique real, that is, for all finite $x \in H\mathbb{R}$, there is a unique $r \in \mathbb{R}$ such that the difference between x and $[f_r]_\approx$ is either infinitesimal or $[f_0]_\approx$.*

CHAPTER 10
Goodstein's Theorem

> For nothing worthy proving can be proven,
> Nor yet disproven
>
> Alfred Lord Tennyson

This chapter is devoted to a remarkable theorem proved by R. L. Goodstein in 1944. It is remarkable in many ways. First, it is such a surprising statement that it is hard to believe it is true. Second, while the theorem is entirely about *finite* integers, Goodstein's proof uses *infinite* ordinals. Third, 37 years after Goodstein's proof appeared, L. Kirby and J. Paris proved that the use of infinite sets is actually *necessary*. That is, this is a theorem of arithmetic that can't be proved arithmetically, but *only* by using the extra powers of set theory!

To describe the theorem, we must discuss what it means to write a number in "superbase 2." Take a number, for example, 23. Writing this in base 2:

$$23_{10} = 10111_2,$$

which stands for the sum:

$$2^4 + 2^2 + 2^1 + 2^0.$$

This expression has a "4" in it. In superbase 2 we want to eliminate all numbers greater than 2, so we write 4 also as a sum of powers of 2: $4 = 2^2$, so we have:

$$23 = 2^{(2^2)} + 2^2 + 2^1 + 2^0 \quad \text{or} \quad 2^{(2^2)} + 2^2 + 2 + 1.$$

For a larger number we might have to go further, for example,

$$514 = 2^9 + 2$$
$$= 2^{(2^3+1)} + 2$$
$$= 2^{(2^{(2+1)}+1)} + 2.$$

The same principle applies to superbase 3, superbase 4, etc.
Here is the theorem:

Take any natural number, write it in superbase 2.
Replace all the '2's with '3's. Subtract 1. Rewrite in superbase 3.
Replace all the '3's with '4's. Subtract 1. Rewrite in superbase 4.
⋮

Goodstein says that eventually you will reach the number 0!
If we follow the steps for a small number, 8:

(start)	$2^{(2+1)} = 8$	
	$3^{(3+1)} = 81$ (changing '2's to '3's)	
(after 1 step)	$2 \cdot 3^3 + 2 \cdot 3^2 + 2 \cdot 3 + 2 = 80$	$(-1, \text{rewrite})$
	$2 \cdot 4^4 + 2 \cdot 4^2 + 2 \cdot 4 + 2 = 554$	('3' to '4's)
(after 2 steps)	$2 \cdot 4^4 + 2 \cdot 4^2 + 2 \cdot 4 + 1 = 553$	$(-1, \text{rewrite})$
	$2 \cdot 5^5 + 2 \cdot 5^2 + 2 \cdot 5 + 1 = 6311$	
(after 3 steps)	$2 \cdot 5^5 + 2 \cdot 5^2 + 2 \cdot 5 = 6310$	
	$2 \cdot 6^6 + 2 \cdot 6^2 + 2 \cdot 6 = 93396$	
(after 4 steps)	$2 \cdot 6^6 + 2 \cdot 6^2 + 6 + 5 = 93395$	

and so on.

The next few numbers are: 1647195, 33554571, 774841151, 20000000211, and 570623341475. How can we possibly get to 0?

Definition. For each $n < \omega$, $2 \leqslant n$, we define a function S_n from ω to ω as follows:

$$S_n(0) = 0$$
$$S_n(k \cdot n^t) = k(n+1)^{S_n(t)} \quad \text{if } k < n,$$

and

10. Goodstein's Theorem

$$S_n\left(\sum_{i=0}^{d} k_i \cdot n^i\right) = \sum_{i=0}^{d} S_n(k_i \cdot n^i),$$

where $k_0, \ldots, k_d < n$.

Definition. For each $n < \omega$, $2 \leq n$, we define a function g_n from ω to ω as follows:

$$g_2(m) = S_2(m) - 1$$

and

$$g_{n+1}(m) = S_{n+1}(g_n(m)) - 1, \quad \text{for } 2 \leq n.$$

10.1. Theorem (Goodstein, Kirby, Paris). $\forall m \exists n\, g_n(m) = 0$.

This is actually an extension of Goodstein's original result, due to Kirby and Paris.

PROJECT #43.

(1) Calculate $g_2(11), g_3(11), \ldots, g_8(11)$.
(2) Find the smallest n such that $g_n(3) = 0$.
(3) Estimate the smallest n such that $g_n(4) = 0$.

Definition. For ordinals α, β, λ, where λ is a limit ordinal,

$$\alpha^1 = \alpha$$
$$\alpha^{\beta+1} = \alpha^\beta \cdot \alpha$$
$$\alpha^\lambda = \bigcup\{\alpha^\delta | \delta < \lambda\}.$$

Definition. For each $n < \omega$, $2 \leq n$, we define a function f_n from ω to the ordinals as follows:

$$f_n(0) = 0$$
$$f_n(k \cdot n^t) = \omega^{f_n(t)} \cdot k, \quad \text{if } k < n,$$

and

$$f_n\left(\sum_{i=0}^{d} k_i \cdot n^i\right) = \sum_{i=0}^{d} f_n(k_i \cdot n^i),$$

where $k_0, \ldots, k_d < n$.

PROJECT #44. Prove:

10.2. Lemma. *For all* $n, m < \omega$, $f_n(m) = f_{n+1}(S_n(m))$.

10.3. Lemma. *For all* $n, m < \omega$, $f_n(m + 1) > f_n(m)$.

10.4. Lemma. *For all* $n, m < \omega$, *if* $g_n(m) > 0$ *then*
$$f_{n+2}(g_{n+1}(m)) < f_{n+1}(g_n(m)).$$

Prove Goodstein's Theorem.

Why are infinite sets needed to prove Goodstein's Theorem? Set theorists describe arithmetic as consisting of the four Peano's Axioms of Chapter 2. Actually, the induction axiom is only stated to hold for definable sets. This system is called *PA*. Goodstein's proof uses ZF. In the years following Goodstein's proof, set theorists and number theorists searched for a proof that used only PA. Finally in 1981, Kirby and Paris showed that Goodstein's Theorem actually implied the consistency of PA, and thus by Gödel's second Incompleteness Theorem (Chapter 6), PA can never prove Goodstein's Theorem.

This brief explanation passes over quite a bit of intricate mathematics. Goodstein's Theorem is a deep fact of numbers, logic, and sets that links together much of the material in this outline. A key ingredient is the use of nonstandard models of arithmetic—formed from N in the same way that HR was formed from R in the last chapter.

PART TWO
SUGGESTIONS

CHAPTER 1
Logic and Set Theory

PROJECT #1. 1. These are the answers to a few:

(i) T   (iii) F   (v) T. This is true since $b \in c$ is false. It may seem peculiar that $\varphi \to \psi$, which we sometimes read as "φ implies ψ" and sometimes as "if φ then ψ," is automatically true if φ is false. In fact this occurs in ordinary English as well. There is a law where I live, for example, that forbids parking at night during the winter months. This can be expressed more precisely as "If it is between the hours of 12 AM and 7 AM on a morning in December, January, February, or March, then you may not park your car on the street." This is a municipal implication, and I find I can obey it without any effort in June. Since the first part of the implication is false, it doesn't matter where I park (as far as *this* law is concerned).

(viii) T   (xiii) F (for example, b).

2. For (xiii), $t \in s$ is a statement by (1), $\exists t (t \in s)$ is a statement by (2g), and so $\forall s \exists t (t \in s)$ is a statement by (2f).

PROJECT #2. You are really being asked to learn a new language and translate it. In approaching each problem, try to put into English what each part means. For example, let's look at (vii). $\exists z (z \in x)$ tells us that x has at least one member, z. $(w \in x \to w = z)$ tells us that if w belongs to x, then w is equal to z. Since "$\forall w$" appears before this, it is true for all w, and so this really says that z is the *only* member of x. Putting these together, statement (vii) says that x has exactly one element. Thus x must be the set e.

Be careful that you recognize in these problems the difference between $\subseteq$ and $\in$. $e \subseteq b$ but $e \notin b$. $d \in b$ but $\neg d \subseteq b$.

PROJECT #3. Now we translate from English to $\mathscr{L}$, a more difficult task. An example: the pair set axiom. We want to say something about any two sets, so on the outside we should have: $\forall c \forall d$. This axiom says that there exists a set, so next we have $\exists e$. What is the relationship between c, d and e? In fact it is just $e = \{c, d\}$, and we are done:

$$\forall c \forall d \exists e (e = \{c, d\}).$$

For practice, if we didn't have the abbreviation $\{c, d\}$, how could we express this? To say c and d are members of e, write $c \in e \wedge d \in e$. To say that these are the *only* members of e, is to say: whenever f is in e, then f must be either c or d:

$$\forall f (f \in e \rightarrow (f = c \vee f = d)).$$

Putting these together:

$$\forall c \forall d \exists e (c \in e \wedge d \in e \wedge \forall f (f \in e \rightarrow (f = c \vee f = d))).$$

Keep in mind the rules of statement formation (1) and (2) stated earlier. Be sure you can verify (as we did in Project #1) that your answers are legitimate statements.

PROJECT #4. 1.1. Show that $A \cap B$ is a definable subset of A, then use Comprehension. Don't forget to write down in $\mathscr{L}$ (together with constants for sets) the explicit statement φ.

1.2. Use Pair Set first, then Union.

1.3. Use Pair Set.

1.4. This is our first use of Regularity. This axiom seems very strange at first, but it recognizes a fundamental property of sets: that each set, no matter how complicated, is composed of sets which are somehow less complicated. Think of b (in the statement of Regularity) as being the simplest element of d. Then all of b's members are simpler than b (hence not in d). To prove 1.4, apply Regularity to the set $\{A\}$.

1. Logic and Set Theory

One consequence of this theorem is that Russell's paradox no longer haunts us. Actually, the paradox disappears even without using Regularity. Instead, we simply have a proof that $R = \{x | x \notin x\}$ is not a set (because if it were, then we would have $R \in R$ iff $R \notin R$—an impossibility).

1.5. Tricky again—apply Regularity to $\{A, B\}$.

1.6, 1.7. Use Extension.

PROJECT #5. $\{a, b\}$ fails, for example. If $c \neq d$, then we want $\langle c, d \rangle$ and $\langle d, c \rangle$ to be different sets, but $\{c, d\}$ and $\{d, c\}$ are the same set. $\{\{a\}, \{b\}\}$ and $\{a, b, \{a, b\}\}$ fail for similar reasons.

$\{a, \{b\}\}$ fails for a different reason. Suppose $c \neq d$. Then we want $\langle \{c\}, d \rangle$ and $\langle \{d\}, c \rangle$ to be different sets, but if we used the definition: $\langle a, b \rangle = \{a, \{b\}\}$, then they would be the same.

To show that $\{\{a\}, \{a, b\}\}$ succeeds, you must show that whenever $\{\{a\}, \{a, b\}\} = \{\{c\}, \{c, d\}\}$, then $a = c$ and $b = d$.

1.8. Show that $A \times B$ is a definable subset of $P(P(A \cup B))$.

1.9. Show that the range of f is a definable subset of $\bigcup(\bigcup f)$.

1.10. Show that $f \upharpoonright D$ is a definable subset of f.

PROJECT #6. 1.12. Use Comprehension.

1.13. Suppose that $[a]_R \cap [b]_R \neq \varnothing$. Show first that aRb. Remember that to prove two sets x and y equal, you should show: (i) that every member of x is also a member of y, and (ii) that every member of y is also a member of x.

1.14. Use Comprehension.

CHAPTER 2
The Natural Numbers

PROJECT #7. 2.2. Suppose a and b both satisfy the Axiom of Infinity. Apply Regularity to the set $(a\setminus b)\cup(b\setminus a)$.

2.3. (d) Use Theorem 1.5
(e) Apply regularity to $\mathbb{N}\setminus X$.

2.4. Let $A = \{k\in\mathbb{N}\mid\varphi(k)\}$.

PROJECT #8. 2.6. By induction on c: let $\varphi(c)$ be the statement:
$$\forall a \forall b (a +_\mathbb{N} (b +_\mathbb{N} c) = (a +_\mathbb{N} b) +_\mathbb{N} c).$$

2.8. By induction on n: let $\varphi(n)$ be the statement: $0 +_\mathbb{N} n = n$.

PROJECT #9. 2.9. Commutativity turns out to be much trickier than associativity. Try this: first prove $a +_\mathbb{N} 1 = 1 +_\mathbb{N} a$ for all $a\in\mathbb{N}$ (by induction), then prove $a +_\mathbb{N} b = b +_\mathbb{N} a$ from this (also by induction).

PROJECT #10. 2.12. Use induction on p.

2.13. You don't need induction.

2.14. Use induction and 2.11.

PROJECT #11. 2.15. Once again, the commutative law is the most difficult. First prove $0 \cdot_N a = 0$ by induction. Next, start to prove $a \cdot_N b = b \cdot_N a$ using induction on a. In the middle of the proof, you will be assuming $a \cdot_N b = b \cdot_N a$ for all $b \in \mathbb{N}$ and will be trying to prove $S(a) \cdot_N b = b \cdot_N S(a)$ for all $b \in \mathbb{N}$. In order to do this, you will have to use induction *again*, this time on b. This kind of proof is sometimes called a *double induction proof*.

PROJECT #12. 2.18. Use induction on y. Remember that if $z \neq 0$, then $z = S(k)$ for some k. Remember also that $S(y) = y \cup \{y\}$.

2.19. Use 2.18 to help prove irreflexivity. Use induction to prove trichotomy.

CHAPTER 3
The Integers

PROJECT #13. 3.1. As you try to prove transitivity you will realize that you are missing an important fact about $\mathbb{N}$, a cancellation law:

$$\text{If} \quad a +_\mathbb{N} b = a +_\mathbb{N} c \quad \text{then} \quad b = c.$$

Call this fact Lemma 2.20 and prove it. Notice that this rule is like subtraction. It is just what is needed to define $\mathbb{Z}$.

Defining $+_\mathbb{Z}$: and why are we expanding from $\mathbb{N}$ to $\mathbb{Z}$? Basically so that we can subtract. Think of each ordered pair $\langle a, b \rangle$ as being the answer to the question: what is $a - b$? For example, $\langle 3, 2 \rangle$ in $\mathbb{Z}$ is really 1, and $\langle 2, 5 \rangle$ is really -3. With this insight, the problem of deciding what $[\langle a, b \rangle]_\sim +_\mathbb{Z} [\langle c, d \rangle]_\sim$ is, is the problem of finding a $[\langle e, f \rangle]_\sim$ such that $(a - b) + (c - d) = (e - f)$.

What should e and f be? Remember, they must be in $\mathbb{N}$, like a, b, c, and d. In fact, there are many possibilities, but the easiest is: $e = a +_\mathbb{N} c$, $f = b +_\mathbb{N} d$.

We are not done yet, however! We must show that $+_\mathbb{Z}$ is *well-defined*. We have decided that

$$[\langle a, b \rangle]_\sim +_\mathbb{Z} [\langle c, d \rangle]_\sim = [\langle a +_\mathbb{N} c, b +_\mathbb{N} d \rangle]_\sim.$$

What this really means is that given k and t in $\mathbb{Z}$, we compute $k +_\mathbb{Z} t$ by choosing a pair $\langle a, b \rangle \in k$ and a pair $\langle c, d \rangle \in t$ (remember, integers are actually sets of pairs of natural numbers) and forming $[\langle a +_\mathbb{N} c, b +_\mathbb{N} d \rangle]_\sim$. But what if we had chosen *different* pairs $\langle a', d' \rangle \in k$ and $\langle c', d' \rangle \in t$? We now get $[\langle a' +_\mathbb{N} c', b' +_\mathbb{N} d' \rangle]_\sim$. If our addition makes sense, our two answers should be the same, that is, if $\langle a,b \rangle \sim \langle a',b' \rangle$ and $\langle c,d \rangle \sim \langle c',d' \rangle$, then we should have $\langle a +_N c, b +_N d \rangle \sim \langle a' +_N c', b' +_N d' \rangle$. Prove this. This is what is meant when we say $+_Z$ is "well-defined."

Leopold Kronecker (quoted at the start of the chapter) would not have approved of our construction of Z. Most especially he would have objected to the use of infinite sets (each $[\langle a,b \rangle]_\sim \in Z$ is infinite). Toward the end of his career, Kronecker doubted strongly the existence of infinite sets. With a little imagination, the reader can probably see how we could construct a set equivalent to Z in which each member is a finite set.

PROJECT #14. To define 0_Z: choose a particular $[\langle a,b \rangle]_\sim$ which will make it easy to prove Theorem 3.4.

To define $-(a)_Z$: if we think of $\langle c,d \rangle$ as meaning $c - d$, then the negative of this is $d - c$, that is, $-([\langle c,d \rangle]_\sim)_Z = [\langle d,c \rangle]_\sim$. Once again, you must show that this definition is well-defined, that is, if $\langle c,d \rangle \sim \langle c',d' \rangle$, then $\langle d,c \rangle \sim \langle d',c' \rangle$.

To define $<_Z$: the natural definition is $[\langle a,b \rangle]_\sim <_Z [\langle c,d \rangle]_\sim$ iff $a +_N d <_N b +_N c$. To prove this is well-defined, you will need another lemma about $\mathbb{N}$: for all $a, b, c \in \mathbb{N}$, $a <_N b$ iff $a +_N c <_N b +_N c$. Call this Lemma 2.21 and prove it.

3.6. Note that from Lemma 2.21 we know that if $a <_N b$ and $c <_N d$ then $a +_N c <_N b +_N d$ (why?).

PROJECT #15. To define $[\langle a,b \rangle]_\sim \cdot_Z [\langle c,d \rangle]_\sim$, think of $(a - b)(c - d) = ac + bd - bc - ad$. This suggests the answer: $[\langle (a \cdot_N c) +_N (b \cdot_N d), (b \cdot_N c) +_N (a \cdot_N d) \rangle]_\sim$. Proving this is well-defined is arduous. Use the fact that $x = y$ implies $x \cdot_N z = y \cdot_N z$ and use it frequently.

In one sense, $\mathbb{N}$ and $\mathbb{Z}$ are completely different sets. In another, $\mathbb{Z}$ contains something that looks just like $\mathbb{N}$, namely the collection:

$$\{[\langle n,0 \rangle]_\sim \in \mathbb{Z} | n \in \mathbb{N}\}.$$

Without being precise, this set behaves just like $\mathbb{N}$. In this sense, we say that $\mathbb{N}$ is *embedded* in $\mathbb{Z}$.

CHAPTER 4
The Rationals

PROJECT #16. When we expanded from $\mathbb{N}$ to $\mathbb{Z}$, we acquired subtraction (think of $a - b$ as $a +_\mathbb{Z} {}^-(b)_\mathbb{Z}$). We did this (sort of) by forming $\mathbb{Z}$ as the collection of all subtraction problems: each $\langle a,b \rangle$ representing $a - b$. That is how we arrived at the relation $\sim$. $\langle a,b \rangle$ and $\langle c,d \rangle$ should represent the same number iff $a - b = c - d$. Since we can't subtract in $\mathbb{N}$, we rewrote this as $a +_\mathbb{N} d = b +_\mathbb{N} c$.

Now to expand $\mathbb{Z}$ to $\mathbb{Q}$, we wish to acquire division. We can think of $\mathbb{Q}$ as ordered pairs $\langle a,b \rangle$, $a, b \in \mathbb{Z}$, which will represent division problems, a/b. Of course b can't be 0. Again, $\langle a,b \rangle$ and $\langle c,d \rangle$ should represent the same number iff $a/b = c/d$. Since we can't divide in $\mathbb{Z}$, we have to rewrite this.

Definition. We define a relation $\approx$ on $\mathbb{Z} \times (\mathbb{Z} \setminus \{0_\mathbb{Z}\})$ by:

$$\langle a,b \rangle \approx \langle c,d \rangle \quad \text{iff} \quad a \cdot_\mathbb{Z} d = b \cdot_\mathbb{Z} c.$$

Show this is an equivalence relation. It won't be easy! You will need a lemma: for $p, q, r \in \mathbb{Z}$, $q \neq 0_\mathbb{Z}$ and $p \cdot_\mathbb{Z} q = r \cdot_\mathbb{Z} q$, then $p = r$. This is a cancellation law for $\cdot_\mathbb{Z}$. There are many ways to prove this. Try your own, or follow this outline:

(1) First prove a concellation law for $\cdot_\mathbb{N}$.
(2) Assume that $p = [\langle i,j \rangle]_\sim$, $q = [\langle k,s \rangle]_\sim$, $r = [\langle t,u \rangle]_\sim$, $q \neq 0_\mathbb{Z}$, and that $p \cdot_\mathbb{Z} q = r \cdot_\mathbb{Z} q$ and write out what this means. With luck, you will arrive at:

$$((i +_N u) \cdot_N k) +_N ((j +_N t) \cdot_N s)$$
$$= ((i +_N u) \cdot_N s) +_N ((j +_N t) \cdot_N k).$$

Use $<_N$ trichotomy on k and s to finish the proof. Note that $k \neq s$ (why?).

Once again we are using infinite sets to build $\mathbb{Q}$, and once again it could be done (with some difficulty perhaps) with finite sets.

PROJECT #17. How do you add fractions? What is $(a/b) + (c/d)$? It's $(ad + bc)/bd$. You must prove that $b \cdot_Z d$ won't be 0_Z, and don't forget to show $+_Q$ is well-defined.

Definition. $0_Q = [\langle 0_Z, 1_Z \rangle]_\approx$, where $1_Z = [\langle 1, 0 \rangle]_\sim$.

Definition. $^-([\langle p, q \rangle]_\approx)_Q = [\langle ^-(p)_Z, q \rangle]_\approx$.

To show this is well-defined, you may need another lemma, but it is possible to prove it directly.

PROJECT #18.
How do you multiply fractions?
What fraction equals 1?
What fraction times a/b equals 1?
There is a lot of work in this project, but it is not very tricky.

As before, $\mathbb{Z}$ and $\mathbb{Q}$ are completely different, yet $\mathbb{Z}$ can be embedded in $\mathbb{Q}$. Can the reader find a copy of $\mathbb{Z}$ in $\mathbb{Q}$?

PROJECT #19. This one is difficult, primarily because there are many facts about $<_Z$ we never bothered to prove. The best general advice is to try proving 4.5 and make a list of all the facts you need to do the job. You can try this without reading further, or you can use the following more detailed suggestion:

Definition. $x \in \mathbb{Z}$ is *pos* iff $0_Z <_Z x$, and *neg* iff $x <_Z 0_Z$.
Prove:

3.10. The sum of two pos numbers is pos.
3.11. The sum of two neg numbers is neg.
3.12. x is pos iff $^-(x)_Z$ is neg.
3.13. The product of two pos numbers is pos.

4. The Rationals

3.14. The product of a pos and a neg is neg.
3.15. The product of two neg numbers is pos.
4.6. The sum of two positive numbers is positive.
4.7. The sum of two negative numbers is negative.

In proving some of these, you may have to go back (once again) to $\mathbb{N}$ and the definition of $<_\mathbb{N}$.

Further hints: it will be necessary to break up into cases. For example, if $[\langle a,b\rangle]_\approx$ is positive, then consider separately *Case* 1: $0_\mathbb{Z} <_\mathbb{Z} a$, $0_\mathbb{Z} <_\mathbb{Z} b$, and *Case* 2: $a <_\mathbb{Z} 0_\mathbb{Z}$, $b <_\mathbb{Z} 0_\mathbb{Z}$.

CHAPTER 5
The Real Numbers

PROJECT #20. In these studies, we have seen mathematics pulling itself up by its bootstraps. $\mathbb{N}$ was formed from $\emptyset$. Lacking subtraction, we created $\mathbb{Z}$ out of the subtraction problems themselves. Lacking division, we created $\mathbb{Q}$ out of the division problems. What do we lack now? Quite a few numbers really. We can't take square roots, for example, but many other important numbers are missing.

There are "holes" in the rational number line. If we cut the line in two, we get two sets.

$$----------------/----------------$$

It might be that the set on the left has a greatest element:

$$---------------\,]\,(----------------$$
$$\quad\quad\quad\quad\{x \in \mathbb{Q} | x \leq 3\} \quad\quad\quad\quad\quad\quad\quad\quad \{x \in \mathbb{Q} | x > 3\}$$

or it might be that the set on the right has a least element:

$$---------------\,)\,[----------------$$
$$\quad\quad\quad\quad\{x \in \mathbb{Q} | x < 5\} \quad\quad\quad\quad\quad\quad\quad\quad \{x \in \mathbb{Q} | 5 \leq x\}$$

but it *might* happen that neither of these is true.

$$---------------\,)\,(----------------$$
$$\quad\quad\quad\quad\{x \in \mathbb{Q} | x^3 < 2\} \quad\quad\quad\quad\quad\quad\quad\quad \{x \in \mathbb{Q} | x^3 > 2\}$$

This is a hole (in the case above, the hole is where $\sqrt[3]{2}$ should go). Every time we cut the line, there should be a number, either on the left or the right. What we do is to create $\mathbb{R}$ as the set of cuts (schnitt is German for "cut"). Each cut is represented by the set on the left. We include property (2) of schnitt so that the sets $\{x \in \mathbb{Q} | x < 3\}$ and $\{x \in \mathbb{Q} | x \leq 3\}$ don't represent two different numbers.

There is something very satisfying in this approach. We are always creating new numbers systems out of "unsolved problems." -5 is the unsolved problem: what number when added to 7 yields 2? $2/3$ is the unsolved problem: what number when multiplied times 3 yields 2? Finally, $\sqrt[3]{2}$ is the unsolved problem: goes in the hole between $\{x \in \mathbb{Q} | x^3 < 2\}$ and $\{x \in \mathbb{Q} | x^3 > 2\}$?

Now to define $<_\mathbb{R}$: when is one schnitt r less than another, s?

$$\text{———————)———————————}$$
$$r$$

$$\text{——————————————)—————————}$$
$$s$$

Clearly, when $r \subseteq s$. Since we don't want $r = s$, we should insist: $r \subsetneq s$.

5.3. Let $r = \cup X$. Show that r is a schnitt. Show it is the least upper bound of X.

This was not the only way to construct the real numbers. Another, more familiar method, is to use infinite decimal expansions. Yet another is to use equivalence classes of Cauchy sequences. Notice that *all* of these approaches require infinite sets. Kronecker (see the suggestions for Project #13) recognized this and mistrusted the real numbers as a consequence. His declaration that π does not exist is frequently quoted.

PROJECT #21. **5.4.** Most of the four characteristics of a schnitt are easy to verify if you have a few lemmas about $<_\mathbb{Q}$. You will probably need:

(a) $x, y \in \mathbb{Q}$ implies $^-(x +_\mathbb{Q} y)_\mathbb{Q} = {}^-(x)_\mathbb{Q} +_\mathbb{Q} {}^-(y)_\mathbb{Q}$ and $^-(^-(x)_\mathbb{Q})_\mathbb{Q} = x$.
(b) For $w, x, y \in \mathbb{Q}$, $w <_\mathbb{Q} x$ and $y <_\mathbb{Q} z$ imply $w +_\mathbb{Q} y <_\mathbb{Q} x +_\mathbb{Q} z$. and to prove this, use:
(c) Lemma 4.6 (see Project #19).

You may need others, depending on the route you take! Keep in mind this picture of a schnitt:

$$\text{————————————)—————————————}$$
$$\text{(no largest element)}\text{(\textit{may} have a smallest element)}$$

5. The Real Numbers

My teacher would always tell me: "Follow your nose." All very well if you have a good nose! Still, the advice has some meaning. For example, to show that if $a <_\mathbb{Q} b$ and $b \in r +_\mathbb{R} s$ then $a \in r +_\mathbb{R} s$, we follow our noses. What do we know? We know $b \in r +_\mathbb{R} s$, so $b = x +_\mathbb{Q} y$ for some $x \in r$, $y \in s$. What do we want to show? We want $a = x' +_\mathbb{Q} y'$ for some $x' \in r$, $y' \in s$. So our job is to find two numbers x' and y' which add up to a. They should be smaller than x and y, and if $x' \leqslant x$ and $y' \leqslant y$, then we will automatically have $x' \in r$, $y' \in s$.

When in doubt, try the simplest approach first—if it doesn't work, find out why. Here, the simplest approach is to let $x' = x$ and just make y' be smaller than y. Let y' be smaller than y by the same amount that a is smaller than b, i.e., let $y' = y +_\mathbb{Q} a +_\mathbb{Q} {}^-(b)_\mathbb{Q}$. In fact, this works.

PROJECT #22. 5.5. For $r, s, t \in \mathbb{R}$, it is helpful to look at $\{x \in \mathbb{Q} \mid x = a +_\mathbb{Q} b +_\mathbb{Q} c, a \in r, b \in s, c \in t\}$.

Definition. $0_\mathbb{R} = \{x \in \mathbb{Q} \mid x <_\mathbb{Q} 0_\mathbb{Q}\}$.

Why is this a schnitt? To show $q \in 0_\mathbb{R}$ is not the greatest element, consider $p = q \cdot_\mathbb{Q} (1/(1_\mathbb{Q} +_\mathbb{Q} 1_\mathbb{Q}))_\mathbb{Q}$. Note that $q <_\mathbb{Q} 0_\mathbb{Q}$ iff ${}^-(q)_\mathbb{Q}$ is positive. To show that ${}^-(p)_\mathbb{Q}$ is positive, consider $q +_\mathbb{Q} {}^-(p)_\mathbb{Q} +_\mathbb{Q} {}^-(p)_\mathbb{Q}$ and use 4.7 (Project #19).

5.6. Follow your nose! It is not hard to show $r +_\mathbb{R} 0_\mathbb{R} \subseteq r$. For the other direction, if $r +_\mathbb{R} 0_\mathbb{R} <_\mathbb{R} r$, then there is a y:

```
------------------)------ y ------)----------
        r+_Q 0_Q                 r
```

Use the fact that r has an element z greater than y, and that $y +_\mathbb{Q} {}^-(z)_\mathbb{Q} \in 0_\mathbb{R}$.

PROJECT #23. Suppose we are given r

```
-------------------------)----------------
                   r
```

and we wish to find the negative of r. Our first thought is to flip this

```
                                     x
-------------------------)-------·--------
                   r                Q\r
```

over, that is, take $\{x \in \mathbb{Q} \mid {}^-(x)_\mathbb{Q} \in \mathbb{Q} \setminus r\}$:

$$\underset{-(r)_{\mathbb{R}}}{\underline{\hspace{2cm}\cdot\overset{-(x)_{\mathbb{Q}}}{\underline{\hspace{2cm}}})\underline{\hspace{6cm}}}}$$

The only problem is that $\mathbb{Q}\backslash r$ may have a *least* element

$$\underline{\hspace{7cm}})[\underline{\hspace{3cm}}$$

which would mean that $^{-}(r)_{\mathbb{R}}$ would have a *greatest* element.

$$\underline{\hspace{4cm}}](\underline{\hspace{6cm}}$$

What should we do?

To define $r \cdot_{\mathbb{R}} s$, first consider the case where $0_{\mathbb{R}} <_{\mathbb{R}} r, s$. Use this to complete the definition.

Once again, $\mathbb{Q}$ can be embedded in $\mathbb{R}$. Can the reader find a copy of $\mathbb{Q}$ in $\mathbb{R}$?

CHAPTER 6
The Ordinals

PROJECT #24. We meet here yet another of the many faces of induction. Under ordinary circumstances the following principles (on any linearly ordered set) are the same:

(1) $<$ is a well-ordering
(2) every nonempty set has a least element (least with respect to $<$)
(3) there are no infinite descending chains $a_1 > a_2 > a_3 > \cdots$.
(4) (Induction) If a property P is true for the first element, and if whenever P holds for all x below a given z it must also hold for z, then P holds for all elements.

6.2. Before we start proving sets ordinals, notice that to prove (1), we don't have to show irreflexivity (why?). Furthermore, we don't have to show that every subset has a least element, because this follows from Regularity. Prove:

6.2a. Lemma. *If A is a set, then A has an $\in$-least element.*

—which gives us:

6.2b. Corollary. *X is an ordinal iff*

(1′) $\in$ *satisfies transitivity*
(2′) $\in$ *satisfies trichotomy*
(3′) *if $a \in b \in X$ then $a \in X$.*

Now for $\mathbb{N}$, show that $<_\mathbb{N}$ and $\in$ are the same to verify (1′) and (2′) (you have already done half of this).

By the way, it is very easy to confuse (1′) and (3′). To make matters worse, the property (3′) is usually expressed by saying that X (a set, not a relation) is *transitive*. They are not the same. For fun, find a set satisfying (1′) but not (3′), then find a set satisfying (3′) but not (1′).

6.3, 6.4. You can grind these out, but often there are also ways of shortening the proofs. Try for clean, direct arguments.

PROJECT #25. 6.5. One way is easy. For the other, show $\alpha = $ the least element of $\beta \setminus \alpha$.

6.6. What does this mean? The collection of all ordinals may not be a set (in fact it isn't). If you look at the definition of a linear ordering on a *set* (pp. 17–18), you will see that the definition makes sense in this context as well, but it is a good idea to write it out.

You will need to use Theorem 6.5. In trichotomy, if $\alpha \neq \beta$, then either $\alpha \setminus \beta$ or $\beta \setminus \alpha$ will not be empty. If, for example, $\alpha \setminus \beta \neq \varnothing$, show that β is its least element.

6.8, 6.9. Having gone this far, you will not be stumped here. It is worth the effort, however, to find the nicest possible proofs. Mathematics is something of an art, and every mathematician (and student of mathematics) is something of an artist.

6.10. In the early days of set theory, the "set of all ordinals" gave rise to a paradox nearly as famous as Russell's. The *Burali–Forti* paradox was that the set of all ordinals, Ω, was well-ordered, hence it had an order-type (Theorem 6.13), α. Then $\alpha \in \Omega$, and since $\alpha + 1$ is also an ordinal, $\alpha + 1 \in \Omega$, but α is the "length" of Ω, and if $\alpha + 1 \in \Omega$, then $\alpha + 1 < \alpha$—impossible.

Our conclusion, of course, is that Ω is not a set. You do not have to appeal to 6.13 to prove this, 6.8 and 6.9 will suffice.

PROJECT #26. 6.12. The words "transfinite induction" are enough to send shudders through many a professional mathematician. On the other hand, the principle is really quite simple to a set theorist. As noted in the suggestion for Project #24, this is nothing more than the fact that $\in$ well-orders the ordinals.

We will be using this principle later, and it will be just like

6. The Ordinals

induction on $\mathbb{N}$ with one exception. On $\mathbb{N}$ it was enough to show:

(1) $\varphi(0)$, and
(2) for all n, if $\varphi(n)$ then $\varphi(n +_\mathbb{N} 1)$.

This suffices because every $n \in \mathbb{N}$ is either 0 or $n +_\mathbb{N} 1$ for some n. This is not true in transfinite induction because of limit ordinals. We will then have to prove:

(3) if $\varphi(\alpha)$ is true for all $\alpha \in \lambda$, λ is a limit ordinal, then $\varphi(\lambda)$ is true.

In practice, (1) and (3) are usually easy, and (2) will provide the only difficulties.

PROJECT #27. First, a word about the proof of Theorem 6.13: what is really going on here is that we construct *order-isomorphisms*. For each well-ordered set X, we show that we can construct a function f from a unique ordinal to X, such that f is one-to-one, onto, and preserves the ordering, i.e., $\alpha \in \beta$ iff $f(\alpha) < f(\beta)$. This ordinal then, has exactly the same *length* as X. The two sets can be lined up and matched to each other, without reordering them, and with all elements corresponding. This idea is crucial in the definitions of $+_o$ and $\cdot_o$.

To understand $+_o$, imagine adding $\omega +_o 1$. We take a sequence of '□'s of length ω:

$$\square \ \square \ \square \ \cdots$$

and one of length 1:

$$\square$$

and place the second after the first:

$$(\ \square \ \underset{\omega}{\square} \ \square \ \cdots \ )(\ \underset{1}{\square} \ ).$$

The result is:

$$\square \ \square \ \square \ \cdots \ \square$$

which is the next ordinal after ω, or $S(\omega)$. Now suppose we add $2 +_o \omega$. We take a sequence of '□'s of length 2:

$$\square \ \square$$

and one of length ω:

$$\square \ \square \ \square \ \cdots$$

and place the second after the first:

$$(\ \square\ \square\ )(\ \square\ \square\ \square\ \cdots\ ).$$

The result is:

$$\square\ \square\ \square\ \square\ \square\ \cdots$$

a string of boxes of length ω! If this seems odd to you, imagine that we have a gigantic measuring device to measuring the order-types of well-orderings. It consists of an infinite sequence of slots:

To measure a sequence of '$\square$'s, we place the boxes in sequence into the slots until all the boxes are used up. We then read the number under the *first empty slot*. To measure the length of $\omega +_o 1$:

$$(\ \square\ \square\ \square\ \cdots\ )(\ \square\ )$$

the first ω '$\square$'s fill up the first ω slots, and the last $\square$ goes in slot ω. The first empty slot is $S(\omega)$. To measure the length of $2 +_o \omega$:

$$(\ \square\ \square\ )(\ \square\ \square\ \square\ \cdots\ )$$

the first two '$\square$'s go in slots 0 and 1. The first $\square$ of ω goes in slot 2, the next in slot 3, and so on. A little thought should convince you that no box will ever be put in slot ω, but every finite slot will be filled, so that ω is the first empty slot.

$\cdot_o$ is computed in a like manner. Suppose we have $\omega \cdot_o 3$. Then we place 3 strings of ω '$\square$'s and place them together.

$$\left[(\ \square\ \square_\omega \square\ \cdots\ )(\ \square\ \square_\omega \square\ \cdots\ )(\ \square\ \square_\omega \square\ \cdots\ )\right].$$

The result is:

$$\square\ \square\ \square\ \cdots\ \square\ \square\ \square\ \cdots\ \square\ \square\ \square\ \cdots$$

also known as: $\omega +_o \omega +_o \omega$.

You may realize by now that $+_o$ is not commutative. $\cdot_o$ is not commutative either, and something odd happens in distributivity too.

CHAPTER 7
The Cardinals

PROJECT #28. 7.1. Remember that to show $\|A\| \leqslant \|B\|$, you only need to find a one-to-one function from A into (not necessarily onto) B. If a further hint is needed, it is helpful to recall George Gamov's story of the hotel with an infinite number of rooms numbered $1, 2, 3, \ldots$. The hotel was full and a traveller arrived needing a place to stay. The manager said there was plenty of room and asked everyone to move to the room with the next higher number, (i.e., the guest in room #76 moves to room #77, etc.). This left room #1 vacant for the traveller.

7.2. For the purposes of this theorem, let us think of $\mathbb{Z}$ as the set: $\{\ldots, -2, -1, 0, 1, 2, \ldots\}$. One way is easy. For the other, return to Gamov's hotel. Now an infinite number of new guests arrive. "No problem!" says the manager, and he now asks each guest to move to the room whose number is twice that of the room currently occupied (i.e., the guest in room #76 moves to room #152, etc.). This leaves all the odd-numbered rooms vacant, enough to handle an infinite number of new arrivals!

7.3. For the purposes of this theorem, let us think of $\mathbb{Q}$ as fractions a/b with $a, b \in \mathbb{Z}$ (which in fact it is). One way is easy. The other can be approached in several ways. The problem is solved if we can line up all the rationals in a line:

$$r_1, r_2, r_3, \ldots$$

which includes every rational number. Start by listing all fractions that can be formed using the numbers $\{-1, 0, 1\}$, then include the numbers that can be formed if $\{-2, 2\}$ are also used, ...

PROJECT #29. 7.4. It looks as though this theorem should be easy, but it's not. Suppose we have f mapping A into B and g mapping B into A, both one-to-one. The problem is that neither may be onto.

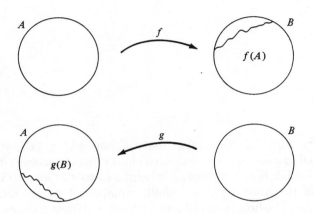

For any $X \subseteq A$, $Y \subseteq B$, we will write:
$$f(X) \text{ for } \{f(x) | x \in X\},$$
and
$$g(Y) \text{ for } \{g(y) | y \in Y\}.$$

The method we will use is to split A up into two sets Z and $A \backslash Z$, and split B up into $f(Z)$ and $B \backslash f(Z)$.

f maps Z one-to-one onto $f(Z)$. If we choose Z properly, g will map $B \backslash f(Z)$ one-to-one onto $A \backslash Z$, and then we can construct a single

7. The Cardinal Numbers

function h by defining:
$$h(x) = y \quad \text{if} \quad x \in Z \quad \text{and} \quad f(x) = y,$$
and
$$h(x) = z \quad \text{if} \quad x \in A \setminus Z \quad \text{and} \quad g(z) = x.$$

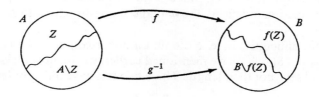

The difficult part is finding an appropriate set Z. We start by defining a function H from $P(A)$ to $P(A)$ by:
$$H(X) = A \setminus g(B \setminus f(X)).$$

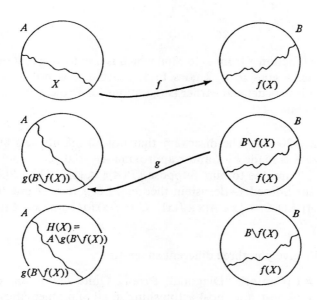

Let $Z = \{a \in A \mid \text{for some } X \subseteq A, a \in X \text{ and } X \subseteq H(X)\}$. Prove:
(1) $X \subseteq Y$ implies $H(X) \subseteq H(Y)$.
(2) $Z \subseteq H(Z)$.

(3) g maps $B \backslash f(Z)$ into $A \backslash Z$.
[Hint: suppose $a \in B \backslash f(Z)$ and $g(a) = b \in Z$. Then $b \in X \subseteq H(X)$ for some $X \subseteq A$. $b \in H(Z)$ (why?). Show this is impossible.]

(4) g maps $B \backslash f(Z)$ *onto* $A \backslash Z$.
[Hint: suppose $b \in A \backslash Z$. Since $b \notin Z$, $Z \cup \{b\}$ is not contained in $H(Z \cup \{b\})$. Since $Z \subseteq H(Z) \subseteq H(Z \cup \{b\})$, we must have $b \notin H(Z \cup \{b\})$. Show then that $b = g(a)$ for some $a \in B \backslash f(Z)$.]

(5) The function h defined above maps A one-to-one onto B.

7.5. No difficulties here. Note that another consequence of 7.4 is that $\|\omega\| = \|S(\omega)\|$. As we mentioned earlier, we are now measuring the size of sets, not their length. These two sets have different *lengths*:

$$\omega: 0 \ 1 \ 2 \ 3 \ 4 \ \cdots$$

$$S(\omega): 0 \ 1 \ 2 \ 3 \ 4 \ \cdots \ \omega.$$

They are *not* order-isomorphic. They are, however, the same *size* because if we rearrange $S(\omega)$:

$$\omega: 0 \ 1 \ 2 \ 3 \ 4 \ \cdots$$

$$S(\omega): \omega \ 0 \ 1 \ 2 \ 3 \ \cdots.$$

We get a function f from ω to $S(\omega)$ which is one-to-one and onto (0 goes to ω, 1 goes to 0, 2 goes to 1, ...) but it is not an order-isomorphism as defined earlier (suggestions for Project #27), since $f(1) \in f(0)$ but $1 \notin 0$.

PROJECT #30. 7.6. The discovery that not all infinite sets are the same size was made by Cantor approximately 100 years ago. Both 7.6 and 7.7 are due to him. Suppose $\|P(A)\| \leq \|A\|$. The $\|P(A)\| = \|A\|$ by the Shroeder–Bernstein theorem. Let f map A one-to-one onto $P(A)$. Let $C = \{x \in A \mid x \notin f(x)\}$. $C = f(x)$ for some $x \in A$ (why?). Is $x \in f(x)$?

7.7. We give you three different suggestions.

PROOF #1 (Cantor's "Diagonal" Proof). Quite simply one of the loveliest, subtlest, and most astonishing in all of mathematics. This author first saw it as a young student and decided that if this is what mathematicians could do, he wanted to be a mathematician.

Let us assume the decimal representation of $\mathbb{R}$. Since $\|\omega\| \leq \|\mathbb{R}\|$ is clearly true, then if $\|\mathbb{R}\| \leq \|\omega\|$ we would have $\|\omega\| = \|\mathbb{R}\|$ by the

7. The Cardinal Numbers

Shroeder–Bernstein Theorem. This would mean that there would be a one-to-one correspondence between real numbers and positive integers. Suppose we had such a correspondence:

$$0 \leftrightarrow r_0$$
$$1 \leftrightarrow r_1$$
$$2 \leftrightarrow r_2$$
$$\vdots$$

in which *all* real numbers are listed. We will show this is impossible. Take the following to illustrate the process:

$$0 \leftrightarrow \quad 23.718044382\ldots$$
$$1 \leftrightarrow \quad -.032196149\ldots$$
$$2 \leftrightarrow \quad 326.461988823\ldots$$
$$3 \leftrightarrow \quad 7.102470065\ldots$$
$$4 \leftrightarrow -13.320795213\ldots$$
$$5 \leftrightarrow \quad 206.423630912\ldots$$
$$\vdots$$

and suppose this list contains *all* the real numbers. Now consider the digits:

$$0 \leftrightarrow \quad 23.\mathbf{7}18044382\ldots$$
$$1 \leftrightarrow \quad -.0\mathbf{3}2196149\ldots$$
$$2 \leftrightarrow \quad 326.46\mathbf{1}988823\ldots$$
$$3 \leftrightarrow \quad 7.102\mathbf{4}70065\ldots$$
$$4 \leftrightarrow -13.3207\mathbf{9}5213\ldots$$
$$5 \leftrightarrow \quad 206.42363\mathbf{0}912\ldots$$
$$\vdots$$

Form a new number by adding 1 to each of the over-size digits above (unless the digit is "9," in which case we subtract 1). This gives us

$$r = .842581\ldots.$$

Show that this number does not appear in the list, contradicting our assumption.

PROOF #2. Since $\|\omega\| < \|P(\omega)\|$, show $\|P(\omega)\| \leq \|\mathbb{R}\|$, then show that if $\|A\| < \|B\| \leq \|C\|$ then $\|A\| < \|C\|$. For the first part, let f map $P(\omega)$ to $\mathbb{R}$ by:

$$f(K) = \sum_{n \in K} 10^{-n}, \quad \text{for any} \quad K \in P(\omega).$$

PROOF #3. We proceed as in Proof #1, but we use schnitts instead of decimal expansions. Suppose we have a one-to-one correspondence:

$$0 \leftrightarrow r_0$$
$$1 \leftrightarrow r_1$$
$$2 \leftrightarrow r_2$$
$$\vdots$$

Where each r_i is a schnitt. We will find a schnitt r that was left out. We do this by forming two sequences of schnitts:

$$s_0 \leq s_1 \leq s_2 \leq \cdots \quad \cdots \leq t_2 \leq t_1 \leq t_0$$

—where all the s_i's are less than all the t_j's, and then make r the least upper bound of the s_i's. Note that r will always be between s_i and t_i. □

Step Zero. Pick a schnitt s_0 such that $r_0 < s_0$. Since $s_0 \leq r$, this will mean $r_0 \neq r$. At the same time, pick a schnitt t_0 such that $s_0 < t_0$.

```
----------+-----+---------------+----------
          r₀    s₀              t₀
```

Step One. Look at r_1. If $r_1 < s_0$

```
-----+----------+---------------+----------
     r₁         s₀              t₀
```

or $t_0 < r_1$,

```
------------------+-------------+--+------
                  s₀            t₀ r₁
```

then we know r will be different from r_1, so we let $s_1 = s_0$, $t_1 = t_0$ and move on to the next step.

Otherwise, if $s_0 \leq r_1 \leq t_0$,

7. The Cardinal Numbers

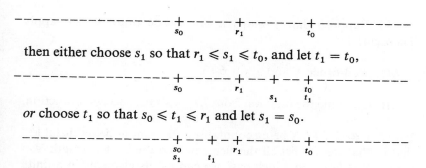

then either choose s_1 so that $r_1 \leqslant s_1 \leqslant t_0$, and let $t_1 = t_0$,

or choose t_1 so that $s_0 \leqslant t_1 \leqslant r_1$ and let $s_1 = s_0$.

In either case, r_1 will not be between s_1 and t_1.

Step two is just like step one, except that we deal with r_2 this time. At each step n, we make sure that the r we eventually form does not equal r_n.

$$s_0 \leqslant s_1 \leqslant s_2 \leqslant \cdots \leqslant r \leqslant \cdots \leqslant t_2 \leqslant t_1 \leqslant t_0.$$

Note: this proof *appears* to use the Axiom of Choice, since we are choosing an infinite number of schnitts. The proof can be changed so that we choose members of $\mathbb{Q}$ (for example, instead of choosing s_0 directly, choose $q_0 \in \mathbb{Q}$ and let

$$s_0 = \{x \in \mathbb{Q} | x <_\mathbb{Q} q_0\}.$$

We can choose members of $\mathbb{Q}$ since $\mathbb{Q}$ can be well-ordered (Theorem 7.5), and being well-ordered allows us to make choices (Theorem 7.10).

7.8. Let $\alpha = \cup A$. If $\beta \in \alpha$ and $\|\beta\| = \|\alpha\|$, show $\|\gamma\| \leqslant \|\beta\|$ for some γ with $\beta \in \gamma \in A$.

7.9. See the suggestion for 7.5.

PROJECT #31. 7.13. Prove this by induction on n. For the induction step, suppose that f maps $n +_\mathbb{N} 1$ one-to-one onto a proper subset of $n +_\mathbb{N} 1$. Define h on n by:

$$h(k) = f(k) \quad \text{if} \quad f(k) \neq n,$$

and

$$h(k) = f(n) \quad \text{if} \quad f(k) = n.$$

Show that h maps n one-to-one onto a proper subset of n.

7.14. Suppose that f maps n one-to-one onto X and apply Theorem 1.11.

7.15. Well-order X and apply 6.13.

7.16. Combine the function from 7.15 with the successor function.

PROJECT #32. **7.17.** X being countable is the same as being "listable," that is, that we can list all the elements and the list has length $\leq \omega$ (either X is finite, in which case we can list its elements in a finite list, or X is infinite and there is a one-to-one map f from ω to X, and then $f(0), f(1), f(2), \ldots$ is a list of its elements, and the length of the list is ω). Let us list the elements of X going across:

$$x_0 \quad x_1 \quad x_2 \quad x_3 \quad \cdots$$

and then list the members of each x_i below it:

x_0	x_1	x_2	x_3	$\cdots$
$a_{0,0}$	$a_{1,0}$	$a_{2,0}$	$a_{3,0}$	
$a_{0,1}$	$a_{1,1}$	$a_{2,1}$	$a_{3,1}$	
$a_{0,2}$	$a_{1,2}$	$a_{2,2}$	$a_{3,2}$	
$a_{0,3}$	$a_{1,3}$	$a_{2,3}$	$a_{3,3}$	
$\vdots$	$\vdots$	$\vdots$	$\vdots$	

Some of these lists may be finite. Now we list all these elements as:

$$0 \leftrightarrow a_{0,0}$$
$$1 \leftrightarrow a_{0,1}$$
$$2 \leftrightarrow a_{1,0}$$
$$3 \leftrightarrow a_{0,2}$$
$$4 \leftrightarrow a_{1,1}$$
$$5 \leftrightarrow a_{2,0}$$
$$\vdots$$

If any element a appears in two different members of X, it would get listed twice. If this happens, we simply remove all repetitions so that the mapping is one-to-one.

Where is the Axiom of Choice used in this proof?

7. The Cardinal Numbers

7.18. Use Theorem 7.17.

7.19. Use Theorem 7.6, the Well-ordering Theorem, and Theorem 6.13. Given $\aleph$, let γ be the least ordinal such that $\|P(\aleph)\| = \gamma$. Prove that γ is a cardinal and that $\aleph < \gamma$.

Yet another paradox in the early, informal formulation of set theory was *Cantor's paradox*: let K be the set of all cardinals. Then $\cup K$ is a cardinal by Theorem 7.8, and it must be the largest. On the other hand, there is no largest cardinal by Theorem 7.19. How do we resolve this?

PROJECT #33. **7.20.** Define the ordering $<_x$ on X by:

$$a <_x b \quad \text{iff} \quad f(a) \in f(b).$$

7.21. Given a cardinal $\aleph$, let $A = \{\beta | \text{ there is a well-ordering of } \aleph \text{ of order-type } \beta\}$. Use Replacement to prove that A is a set. Prove $\gamma = \cup A$ is a cardinal greater than $\aleph$ as follows: suppose for some $\alpha < \gamma$ that $\|\alpha\| = \|\gamma\|$. Then:

(1) $\alpha \in \beta \in A$ for some β; (why?)
(2) $\|\gamma\| = \|\beta\|$; (why?)
(3) there is a one-to-one function from $\aleph$ onto γ; (why?)
(4) $\gamma \in A$; (why?)
(5) $\gamma +_o 1 \in A$; (why?)
(6) contradiction! (why?)

CHAPTER 8
The Universe

PROJECT #34. 8.2. Our first proof by transfinite induction. Suppose we are given an ordinal α and for all $\beta < \alpha$ the theorem is true, i.e., $x \in y \in V_\beta$ implies $x \in V_\beta$. Suppose $x \in y \in V_\alpha$. You will need to separate the problem into cases:

Case 1. $\alpha = 0$
Case 2. $\alpha = \beta +_o 1$ for some β, and
Case 3. α is a limit ordinal.

8.3. By transfinite induction on α. Suppose the theorem is true for all $\beta < \alpha$. Prove that $V_\delta \subseteq V_\alpha$ for all $\delta < \alpha$.

PROJECT #35. 8.4. A very simple statement, but the proof requires some real work and an introduction to a few standard set-theoretic tricks. The key ingredient, although it is not obvious, is Regularity. This is, in fact, equivalent to Regularity.

(1) In outline, we assume the theorem is false. We then find an x which is not in any V_α but each of whose elements *are* (though each element might be in a different V_α). We then find a single β such that every element of x is in V_β. It will then follow that $x \in V_{\beta+1}$. (Why?)
(2) To find x, first let z be a set not in any V_α. If z does not suffice, that is, if z has elements which are not in any V_α, then we form what is called the *transitive closure* of z: $T(z) =$

$z \cup (\cup z) \cup (\cup \cup z) \cup (\cup \cup \cup z) \cup \ldots$ why is this a set? This is a common application of Replacement. We define a function f with domain ω so that $f(0) = z$, $f(1) = \cup z$, $f(2) = \cup \cup z$, ..., then the range R of f is a set, and $\cup R$ is $T(z)$. Replacement requires that the function be definable. We must find a formula φ so that $\varphi(a, b)$ is true iff $f(a) = b$. The statement is:

"$\exists f(f$ is a function $and\ f(0) = z$
and for all $n \in \omega\ f(n + 1) = \cup f(n)\ and\ f(a) = b)$."

Write this statement in our symbolic language using only approved abbreviations.
(3) We now have $T(z)$. This has the property that if $c \in d \in T(z)$ then $c \in T(z)$. (Why?)
(4) Let $Y = \{s \in T(z) | s$ is in no $V_\alpha\}$. We verify that Y is a set by Comprehension. (How can we express "s is in no V_α" in $\mathscr{L}$?)
(5) Since there are elements of z which are not in any V_α, $Y \neq \emptyset$. By regularity there is a set $x \in Y$ such that $x \cap Y = \emptyset$. x is the set we seek, that is, each $q \in x$ is in some V_α. (Why?)
(6) Finally, we need a β such that each $q \in x$ is in V_β. Define a function g on x by: $g(q) = $ "the least ordinal α such that $q \in V_\alpha$," a definable function. By Replacement, the range R of g is a set. Show that $\beta = \cup R$ is our desired ordinal.

Note: there is actually little mathematical justification for the Axiom of Regularity. Nearly all of conventional mathematics can be accomplished without it. The universe, in its absence, might be different perhaps, but still viable. On the other hand, Regularity imposes a satisfactory order on the hierarchy of sets that would be missed. For this and other reasons, it has become part of standard set theory.

A Second Note: important to this proof and many others is the recognition that certain phrases can be expressed in $\mathscr{L}$ so that the sets or functions they describe are definable. We already have some expertise here, casually allowing "f is a function," "α is an ordinal," etc. As one works in set theory, one learns to recognize what can be expressed and what can't. Indeed, there are areas (such as descriptive set theory) where one must learn to notice quickly such additional subtleties as the arrangement of quantifiers that must be used!

PROJECT #36. 8.5. In one sense this is obvious—how could a finite set ($\|x\| < \omega$) reach all the way up to ω? Still, the way is not clear.

8. The Universe

Given X, $\|X\| = n < \omega$. Since X is well-ordered by ε, there is a one-to-one function f from an ordinal α onto X. Prove:

(1) X is Dedekind finite
(2) $\alpha < \omega$
(3) $\alpha = S(k)$ for some k
(4) $f(k) = \cup X$
(5) $\cup X < \omega$.

8.7. Use Theorem 7.12.

PROJECT #37. 8.9. Suppose this is false. Let α be the least ordinal such that $\|V_\alpha\|$ is not less than κ. There are two cases, α a successor ordinal and α a limit ordinal. To handle the limit ordinal case, let f map κ one-to-one onto V_α. For each $\beta < \alpha$, let $X_\beta = \{\delta \in \kappa | f(\delta) \in V_\beta\}$. Prove:

(1) $\|X_\beta\| < \kappa$ for all $\beta < \alpha$
(2) $\cup X_\beta < \kappa$ for all $\beta < \alpha$
(3) $\cup \{\cup X_\beta | \beta < \alpha\} < \kappa$
(4) this is a contradiction.

PROJECT #38. 8.10. Many of the axioms are fairly easy to handle once we see what we are to prove. For example, Pair Set. We must show that if $x, y \in V_\kappa$ then $\{x, y\} \in V_\kappa$. We are working in ZF, so $\{x, y\}$ *is* a set. We must show it is in V_κ. Use the fact that $V_\kappa = \cup \{V_\alpha | \alpha < \kappa\}$.

Theorem 8.2 is frequently helpful.

For Infinity, prove by transfinite induction that $\alpha \in V_{\alpha +_o 1}$ for all ordinals α.

For Replacement you only have to prove: if f is a function with domain $x \in V_\kappa$ and f is definable in V_κ, then the range of f is a set in V_κ. By Replacement, the range R of f is a set. What we must show is that $R \in V_\kappa$. "Definable in V_κ" implies that for all $a \in x$, $f(a) \in V_\kappa$. Use the trick of (6) in the proof of Theorem 8.4.

We can actually show more, that V_κ is a model for ZFC.

CHAPTER 9
Choice and Infinitesimals

PROJECT #39. 9.1. The idea behind the proof is fairly simple, though the proof is not. Suppose we are given set X to well-order. Choose an element of X. Call that a_1, the first element of X. Next, choose another element of X. Call that a_2, the second element of X. Keep going until all elements of X have been ordered.

There are two difficulties: one, we are making many choices—how can this be done? Two, even if we can make the choices, the process we are describing may take an infinite amount of time (ω steps might not be enough, $\omega +_o \omega$, $\omega \cdot_o \omega$, even $\aleph_1$ steps may not be enough). How do we know it can be done?

Let $Y = P(X)\setminus\{\emptyset\}$ (all nonempty subsets of X). Let h be a choice function for Y, that is, for all $A \in Y, h(A) \in A$. This solves one difficulty, we can use h to make the choices: $a_1 = h(X)$, $a_2 = h(X\setminus\{a_1\})$, $a_3 = h(X\setminus\{a_1, a_2\})$, ...

For the other difficulty, we use the Theorem on Inductive Definitions: let f be the function defined by:

$$f(A) = h(X\setminus A) \quad \text{if} \quad X\setminus A \neq \emptyset.$$

Apply Theorem 6.14 to f to get a new function g. Prove:

(1) g is one-to-one,
(2) the range of g is a set,
(3) the domain of g^{-1} is a set,
(4) g^{-1} and g are functions,

(5) the domain of g is an ordinal, δ,
(6) the range of g on δ is X, and
(7) X is well-ordered.

[Further hint: review the proof of Theorem 6.13.]

9.2. We are given X and a partial ordering $<_x$ on X. Again, the idea behind the proof is primitive, we simply pick elements one by one from X. Pick a_0. If it is not a maximal element, then pick a_1 larger than a_0. If this is not maximal, then pick a larger one, and so on. After we have picked ω-many elements, $a_0 <_x a_1 <_x a_2 <_x \cdots$ we have a chain. By assumption, it is bounded, that is, there is an a_ω above all these,

$$a_0 <_x a_1 <_x a_2 <_x \cdots$$
$$\cdots <_x a_\omega.$$

If a_ω is not maximal, then pick a larger one, and so on. Once again, how can we make choices, and how do we know the process will end?

Using Zermelo's Theorem, well-order X–call this well-ordering $\ll_x$ (it will almost certainly be different from $<_x$). We will use $\ll_x$ to make our choices (to pick a_0, pick the $\ll_x$-least element of X, to pick a_1, pick the $\ll_x$-least element of X which is $>_x a_0$). Now define:

$f(A) = $ the $\ll_x$-least element of X which is $>_x$,
    all elements of A (if there is one).

Again, apply Theorem 6.14 to f to get g. Prove (1)–(5) from the suggestions for 9.1, and prove:

(6′) the range of g is a chain in X.

Let α be the domain of g. Prove:

(7′) α is a successor ordinal $\beta +_o 1$.
(8′) $g(\beta)$ is a maximal element.

PROJECT #40. 9.3. Given X, we are required to find a choice function, i.e., a function f such that if $y \in X$, $y \neq \emptyset$, then $f(y) \in y$. It is no problem to pick a member of y if we are just given one y. Similarly we can easily make any finite number of choices. The difficulty is handling an infinite number.

We solve this by using Zorn's Lemma. Consider the set $P = \{f | f$ is a function with domain $D \subseteq X$, and for all $y \in D$, $y \neq \emptyset$, $f(y) \in y\}$. Each element of P is a "partial" choice function. It makes some

9. Choice and Infinitesimals

choices, but usually not all, that is, for most f, $D \subsetneq X$. Define an ordering $<_p$ on P by:

$$f <_p g \quad \text{iff} \quad f \subsetneq g.$$

Prove:

(1) $<_p$ is a partial ordering.
(2) every chain is bounded, and
(3) a maximal element in P is a complete choice function. [Hint: Suppose f is maximal, $y \in X$, $y \neq \emptyset$. Show that if y is *not* in the domain of f, then there is a $g \in P$ with $f <_p g$.]

9.4. By way of explanation, think of $\mathcal{U}$ as a collection of "big" sets. The properties of the ultrafilters then are:

(a) if two sets are big, then they are so big that their intersection is big,
(b) no finite set is big, and
(c) every set A is either big itself, or its complement is big.

Constructing an ultrafilter means making endless decisions. Given a set A, should we call it big, or should we call $\omega \setminus A$ big? For some sets this is easy, for example, $\{1, 3, 8\}$ cannot be big, so the set $\{0, 2, 4, 5, 6, 9, 10, \ldots\}$ must be big. For other sets such as $\{1, 3, 5, 7, 9, 11, 13, \ldots\}$ there is no clear answer. In fact, it doesn't really matter what sets we choose, so long as the resulting collection satisfies (a), (b) and (c). Is $\{1, 3, 5, 7, \ldots\}$ big? It is in *some* ultrafilters, and in other ultrafilters it's not.

The proof of this theorem is another application of Zorn's Lemma. Let $Q = \{u \subseteq P(\omega) | u \text{ satisfies (a) and (b) of the definition of ultrafilters}\}$. Define an ordering:

$$u <_q v \quad \text{iff} \quad u \subsetneq v.$$

Prove:

(1) $<_q$ is a partial ordering.
(2) Q is not empty.
(3) every chain is bounded, and
(4) A maximal element is an ultrafilter. [Hint: if u is maximal and not an ultrafilter, then there is an $A \subseteq \omega$ such that neither A nor $\omega \setminus A$ is in u. Show that at least one of the two is infinite and has nonempty intersection with all elements of u. Call this set S. Show that $u <_q v = u \cup \{B \cap S | B \in u\} \in Q$.]

PROJECT #41. 9.5. It will help to prove:

9.5a. Lemma. *If $A \in \mathcal{U}$ and $A \subseteq B$ then $B \in \mathcal{U}$*

9.7. Consider $[H]_\approx$, where $H(n) = n$, for all $n \in \omega$.

When "infinitesimals" as very small numbers fell from favor 100 years ago, textbooks, loath to give them up, began to describe them as variables that go to zero. Ironically, that is just what they are here. Fundamentally, they are sequences whose limit (as defined in calculus) is 0.

PROJECT #42. The obvious choice for $+_{\mathsf{HR}}$ is $[f]_\approx +_{\mathsf{HR}} [g]_\approx = [h]_\approx$ where $h(n) = f(n) +_\mathbb{R} g(n)$ for all $n \in \omega$. For exercise, you might show that $+_{\mathsf{HR}}$ is well-defined.

For (1)–(5), it will help to think of H as $[h]_\approx$, K as $[k]_\approx$, I as $[i]_\approx$, J as $[j]_\approx$, where h, k, i, j are in χ. (1) and (2) you will find are not difficult. Consider (3). $I +_{\mathsf{HR}} J$ clearly satisfies (1) and (3) of the definition of infinitesimal; what about (2)? Suppose $r \in \mathbb{R}$ is a positive real. Is it true that $\{n \in \omega | i(n) +_\mathbb{R} j(n) <_\mathbb{R} r\}$ is in $\mathcal{U}$? Since I and J are infinitesimal, we know that $\{n \in \omega | i(n) <_\mathbb{R} r\}$ and $\{n \in \omega | j(n) <_\mathbb{R} r\}$ are in $\mathcal{U}$, but this doesn't seem to help much. [Hint: consider $\{n \in \omega | i(n) <_\mathbb{R} r/2\}$.]

For (6), suppose there were a largest finite hyperreal, $[g]_\approx$. Since $[g]_\approx$ is finite, there is some $r \in \mathbb{R}$ such that $[g]_\approx <_{\mathsf{HR}} [f_r]_\approx$. Is this possible?

For (9) and (10), keep Theorem 9.9 in mind.

Here is a sample of how one uses HR in calculus:

Definition. A function f from $\mathbb{R}$ to $\mathbb{R}$ is *continuous at r* iff for all infinitesimals I, $f(r + I)$ is infinitely close to $f(r)$.

The proper domain of f is $\mathbb{R}$, of course, but every function defined on $\mathbb{R}$ can also be defined on HR (by Theorem 9.9, or by direct computation).

An example: is $y = x^2$ continuous at 2? If I is infinitesimal, then $f(2 + I) = (2 + I)^2 = 4 + 2I + I^2$.

$2I$ and I^2 are infinitesimals (as in (2) above), $2I + I^2$ is infinitesimal (as in (3) above), so $f(2) = 4$ *is* infinitely close to $f(2 + I)$.

9. Choice and Infinitesimals

A Final Note: the use of infinitesimals can make the calculus significantly easier. In general, however, they are best used intuitively. We use real numbers, for example, every day without thinking of schnitts. We add integers frequently without worrying about equivalence classes. Similarly, it is easy to deal with infinitesimals and infinite numbers if we simply rely (as our ancestors did) on a good mental picture of them.

CHAPTER 10
Goodstein's Theorem

PROJECT #43. These functions appear tricky, but the idea is simple. S_n is the function that takes a number written in superbase n and changes all the 'n's to '$n + 1$'s. g_n takes a number and performs n steps in the Goodstein sequence.

(3) As a warm-up, consider these exercises:
    (a) Suppose after 91 steps we are at $13 \cdot 93^0$ (base 93) (never mind how we got here). How many more steps will it take to reach 0?
    (b) Suppose after 35 steps we are at $1 \cdot 37$ (base 37). How many more steps will it take to reach 0?
    (c) Suppose after 43 steps we are at $2 \cdot 45$ (base 45). How many more steps will it take to reach 0?
    (d) Suppose after 1 step we reach $2 \cdot 3^2 + 2 \cdot 3 + 2$ (base 3). How many steps will it take to reduce the right-most term to 0? How many more steps will it take to reduce the middle term to 0?

PROJECT #44. The idea behind this complex-looking proof is not complex. If at step n we replace all 'n's with 'ω's, we get an ordinal. If we do this at each step, we get a sequence of ordinals that is *descending*, that is, the ordinal we get from step 125 is greater than the ordinal we get from step 126, and so on. Since there cannot be an infinite descending sequence of ordinals (because the ordinals are well-ordered), then sequence must end sometime at 0.

The functions f_n are the functions which substitute 'ω's for 'n's.

10.2. For all of these, remember that given a base n, every number $m > 0$ can be written in the form:

$$m = \sum_{i=0}^{d} k_i n^i,$$

where $m > d$ and for each i, $0 \leqslant k_i < n$. This is the unique base n representation. So, if you want to prove something about all integers m by induction, you can proceed by

(1) proving the fact for $m = 0$, and
(2) proving that if the fact is true for all $i \leqslant d$, then it is true for

$$m = \sum_{i=0}^{d} k_i n^i$$

for any $k_0, \ldots, k_d < n$.

10.3. Use induction. First check that the lemma is true for 0. Next, assume it is true for all numbers less than m and write m as:

$$k_d n^d + k_{d-1} n^{d-1} + \cdots + k_1 n + k_0,$$

where $0 \leqslant k_i < n$ for each $i \leqslant d$. Consider

Case 1. $k_0 < n - 1$, and then
Case 2. $k_s < n - 1$ but $k_i = n - 1$ for all $i < s$.

In case 2,

$$m = k_d n^d + \cdots + k_s n^s + (n-1) n^{s-1} + \cdots + (n-1),$$

and

$$m + 1 = k_d n^d + \cdots + (k_s + 1) n^s.$$

Then

$$f_n(m) = f_n(k_d n^d + \cdots + k_{s+1} n^{s+1})$$
$$+ f_n(k_s n^s + (n-1) n^{s-1} + \cdots + (n-1)),$$

while

$$f_n(m + 1) = f_n(k_d n^d + \cdots + k_{s+1} n^{s+1}) + f_n((k_s + 1) n^s).$$

10. Goodstein's Theorem

It will be enough then, to show that $f_n(k_s n^s + (n-1)n^{s-1} + \cdots + (n-1))$ is less that $f_n((k_s + 1)n^s)$.

10.4. You're almost there—just put together the previous lemmas.

10.5. Consider the sequence: $f_3(g_2(m)), f_4(g_3(m)), f_5(g_4(m)), \ldots$.

PART THREE
SOLUTIONS

CHAPTER 1
Logic and Set Theory

PROJECT #1.   1. (i) T   (ii) F   (iii) F   (iv) T   (v) T   (vi) F   (vii) F   (viii) T   (ix) F (c is not a subset of e)   (x) T (e is a subset of a)   (xi) T (d is nonempty)   (xii) F (for example, a)   (xiii) F   (xiv) T (e is a proper subset of a set)   (xv) F (e does not have two distinct elements).

2. (xiv) $i \in e, i \in h, h = e$ are statements by (1).
$(i \in e \rightarrow i \in h)$ a statement by (2d).
$\neg h = e$ is a statement by (2a).
$((i \in e \rightarrow i \in h) \wedge \neg h = e)$ is a statement by (2b).
$\forall i((i \in e \rightarrow i \in h) \wedge \neg h = e)$ is a statement by (2f).
and finally,
$\exists h \forall i((i \in e \rightarrow i \in h) \wedge \neg h = e)$ is a statement by (2g).
(xv) $g \in e, w \in e, g = w$ are statements by (1).
$(g \in \vee w \in e)$ is a statement by (2c).
$\neg g = w$ is a statement by (2a).
$((g \in e \vee w \in e) \wedge \neg g = w)$ is a statement by (2b).
$\exists w((g \in e \vee w \in e) \wedge \neg g = w)$ is a statement by (2g),
and finally,
$\exists g \exists w((g \in e \vee w \in e) \wedge \neg g = w)$ is a statement by (2g).

PROJECT #2. (i) a (ii) f (iii) f (iv) f (v) a (vi) a (vii) e (e is the set with exactly one element)   (viii) d (ix) e (x) f (xi) a (xii) e (xiii) d (xiv) a (xv) a (only a has no proper subsets)

(xvi) a (the only set which does not have an empty set as a member)
(xvii) e (has exactly one proper subset)   (xviii) f (xix) b (xx) d
(d has exactly three distinct elements: a, c, and e).

PROJECT #3.

Extension: $\forall x \forall y (x = y \leftrightarrow \forall z (z \in x \leftrightarrow z \in y))$
Empty Set: $\exists z (z = \{\ \})$
Pair Set: $\forall x \forall y \exists z\, z = \{x, y\}$
Union: $\forall x \exists y \forall z (z \in y \leftrightarrow \exists w (z \in w \land w \in x))$
Power Set: $\forall x \exists y \forall z (z \in y \leftrightarrow z \subseteq x)$
Regularity: $\forall x (x = \{\ \} \lor \exists z (z \in x \land x \cap z = \{\ \}))$.

PROJECT #4. 1.1. By the Pair Set axiom, $\{A, B\}$ is a set. By the Union axiom, $\cup \{A, B\} = A \cup B$ is a set.

1.2. $A \cap B = \{x \in A \mid \varphi(x)\}$ where $\varphi(x)$ is the statement: $x \in B$. Hence $A \cap B$ is a set by Comprehension.

1.3. By the Pair Set axiom, $\{A, A\} = \{A\}$ is a set.

1.4. By Regularity, there is an $x \in \{A\}$ such that $x \cap \{A\} = \emptyset$. x must be A, so $A \cap \{A\} = \emptyset$. In particular, $A \notin A$.

1.5. By Regularity there is an $x \in \{A, B\}$ such that $x \cap \{A, B\} = \emptyset$. Since $A \in B$, x cannot be B, hence $x = A$ and $A \cap \{A, B\} = \emptyset$. This implies $B \notin A$.

1.6. If two sets are both empty, then they have the same members, hence by Extension, they are equal.

1.7. Similar to 1.6

PROJECT #5. Picking up where the suggestions leave off, suppose $\{\{a\}, \{a, b\}\} = \{\{c\}, \{c, d\}\}$. Then either $\{a\} = \{c\}$ or $\{a\} = \{c, d\}$. If $\{a\} = \{c\}$, then $a = c$ and we must have $\{a, b\} = \{c, d\}$ which leads to $b = d$. Suppose, however, that $\{a\} = \{c, d\}$. Then $a = c = d$. Further, $\{a, b\}$ must be $\{c, d\} = \{c, c\} = \{c\}$, so $a = b = c$. In all cases we find $a = c$ and $b = d$.

1.8. For $a \in A$, $b \in B$, $\{a\}$, $\{a, b\} \in P(A \cup B)$, so $\langle a, b \rangle = \{\{a\}, \{a, b\}\} \subseteq P(A \cup B)$, or $\langle a, b \rangle \in P(P(A \cup B))$. Then $A \times B =$

1. Logic and Set Theory

$\{x \in P(P(A \cup B)) | \varphi(x)\}$ where $\varphi(x)$ is: $\exists a \exists b (x = \langle a, b \rangle \land a \in A \land b \in B)$ (we can introduce $\langle a, b \rangle$ to $\mathscr{L}$ as another abbreviation). Then $A \times B$ is a set by Comprehension.

1.9. If $\langle a, b \rangle = \{\{a\}, \{a, b\}\} \in f$, then $\{a, b\} \in \cup f$ (since $\{a, b\} \in \langle a, b \rangle \in f$) and $b \in \cup(\cup f)$ (since $b \in \{a, b\} \in \cup f$). Then the range of $f = \{x \in \cup(\cup f) | \varphi(x)\}$ where $\varphi(x)$ is $\exists y \exists z (z = \langle y, x \rangle \land z \in f)$, hence it is a set by Comprehension.

1.10. $f \upharpoonright D = \{x \in f | \exists y \exists z (\langle y, z \rangle = x \land y \in D)\}$, a set by Comprehension.

1.11. f^{-1} is a set by Comprehension. It is a function since if $\langle a, b \rangle$, $\langle a, c \rangle \in f^{-1}$ then $\langle b, a \rangle, \langle c, a \rangle \in f$, and so $b = c$ since f is one-to-one. Similarly, f^{-1} is one-to-one because f is a function.

PROJECT #6. 1.12. $[a]_R = \{x \in A | \langle x, a \rangle \in R\}$

1.13. If $[a]_R \cap [b]_R \neq \emptyset$, let x be in the intersection. Then aRx and bRx, so xRb by symmetry and then aRb by transitivity. It follows that any $y \in [a]_R$ is also in $[b]_R$ as follows: $y \in [a]_R$ implies aRy so yRa, then using aRb, yRb, so bRy and then $y \in [b]_R$. The other direction is proved in the same way, so $[a]_R = [b]_R$.

1.14. Each $[a]_R$ is in $P(A)$, so the collection of equivalence classes is a subset of $P(A)$. In fact, it is: $\{x \in P(A) | \exists a \forall y (y \in x \leftrightarrow aRy)\}$.

CHAPTER 2
The Natural Numbers

PROJECT #7. 2.1. True by 1.3 and Union.

2.2. If $x \neq y$ then $A = (x\backslash y) \cup (y\backslash x) \neq \emptyset$. By Regularity, let $a \in A$ be such that $a \cap A = \emptyset$. Suppose $a \in x\backslash y$. Then $a \neq \emptyset$, since $\emptyset \in y$, so $a = S(z)$ for some $z \in x$. Since $z \in a$, $z \notin A$. Since $z \notin A$ and $z \in x$, $z \in y$. Since $z \in y$, $a = S(z) \in y$, contradicting $a \in x\backslash y$. A similar contradiction follows if $a \in y\backslash x$.

$0 = \{\ \}$ $1 = \{\{\ \}\}$ $2 = \{\{\ \},\{\{\ \}\}\}$
$3 = \{\{\ \},\{\{\ \}\},\{\{\ \},\{\{\ \}\}\}\}$
$4 = \{\{\ \},\{\{\ \}\},\{\{\ \},\{\{\ \}\}\},\{\{\ \},\{\{\ \}\},\{\{\ \},\{\{\ \}\}\}\}\}$.

2.3(a), (b). By definition of $\mathbb{N}$. (c) $x \in S(x)$, so $S(x) \neq \emptyset = 0$. (d) Clearly $x = y \to S(x) = S(y)$. On the other hand, if $S(x) = S(y)$, then $x \cup \{x\} = y \cup \{y\}$. If $x \neq y$, then $x \in y$ and $y \in x$, violating Theorem 1.5. (e) If $A \neq \mathbb{N}$, then by Regularity, there is an $x \in \mathbb{N}\backslash A$, $x \cap \mathbb{N}\backslash A = \emptyset$. Since $x \neq 0$, $x = S(k)$ for some $k \in \mathbb{N}$. As $k \in x$, $k \notin \mathbb{N}\backslash A$, so $k \in A$. By hypothesis then, $x = S(k) \in A$, a contradiction.

2.4. Let $A = \{x \in \mathbb{N} | \varphi(x)\}$. $0 \in A$ and whenever $k \in A$, $S(k) \in A$, so $A = \mathbb{N}$ by Theorem 2.3 (e), therefore $\varphi(x)$ is true about all $x \in \mathbb{N}$.

PROJECT #8. 2.6. Let $\varphi(c)$ be the statement: $\forall a \forall b (a +_\mathbb{N} (b +_\mathbb{N} c) = (a +_\mathbb{N} b) +_\mathbb{N} c)$. To show $\varphi(c)$ is true for all c (associativity), we use induction (Theorem 2.4). First, $\varphi(0)$ is true, since $a +_\mathbb{N} (b +_\mathbb{N} 0) = a +_\mathbb{N} b = (a +_\mathbb{N} b) +_\mathbb{N} 0$. Next, suppose $\varphi(k)$. Then

$$a +_\mathbb{N} (b +_\mathbb{N} S(k)) = a +_\mathbb{N} S(b +_\mathbb{N} k)$$
$$= S(a +_\mathbb{N} (b +_\mathbb{N} k))$$
$$= S((a +_\mathbb{N} b) +_\mathbb{N} k) \quad \text{(since } \varphi(k))$$
$$= (a +_\mathbb{N} b) +_\mathbb{N} S(k),$$

hence $\varphi(S(k))$.

We conclude that $\varphi(k)$ is true for all k.

2.7. $n +_\mathbb{N} 1 = n +_\mathbb{N} S(0) = S(n +_\mathbb{N} 0) = S(n)$.

2.8. Let $\varphi(n)$ be the statement: $0 +_\mathbb{N} n = n$. $\varphi(0)$ is true since $0 +_\mathbb{N} 0 = 0$. If $\varphi(n)$ holds, then $0 +_\mathbb{N} S(n) = S(0 +_\mathbb{N} n) = S(n)$, so $\varphi(S(n))$ holds. By induction, $\varphi(n)$ is true for all $n \in \mathbb{N}$.

PROJECT #9. 2.9. We begin by proving $a +_\mathbb{N} 1 = 1 +_\mathbb{N} a$ by induction on a. First, $0 +_\mathbb{N} 1 = 1 +_\mathbb{N} 0$ by Theorem 2.8. Next suppose $a +_\mathbb{N} 1 = 1 +_\mathbb{N} a$. Then

$$S(a) +_\mathbb{N} 1 = S(S(a)) \quad \text{(by Theorem 2.7)}$$
$$= S(a +_\mathbb{N} 1)$$
$$= S(1 +_\mathbb{N} a)$$
$$= 1 +_\mathbb{N} S(a).$$

By induction then, $a +_\mathbb{N} 1 = 1 +_\mathbb{N} a$ for all $a \in \mathbb{N}$.

Now we prove $a +_\mathbb{N} b = b +_\mathbb{N} a$ by induction on b. First, $a +_\mathbb{N} 0 = 0 +_\mathbb{N} a$, by Theorem 2.8. Now if $a +_\mathbb{N} b = b +_\mathbb{N} a$, then

$$a +_\mathbb{N} S(b) = a +_\mathbb{N} (b +_\mathbb{N} 1)$$
$$= (a +_\mathbb{N} b) +_\mathbb{N} 1$$
$$= (b +_\mathbb{N} a) +_\mathbb{N} 1$$
$$= b +_\mathbb{N} (a +_\mathbb{N} 1)$$
$$= b +_\mathbb{N} (1 +_\mathbb{N} a)$$
$$= (b +_\mathbb{N} 1) +_\mathbb{N} a$$
$$= S(b) +_\mathbb{N} a.$$

We conclude that $a +_\mathbb{N} b = b +_\mathbb{N} a$ for all $a, b \in \mathbb{N}$.

2. The Natural Numbers

2.10. $2 +_N 2 = 2 +_N S(1) = S(2 +_N 1) = S(S(2)) = S(3) = 4$.

PROJECT #10. 2.12. We prove $n \cdot_N (m +_N p) = (n \cdot_N m) +_N (n \cdot_N p)$ for all $n, m, p \in \mathbb{N}$ by induction on p. First, $n \cdot_N (m +_N 0) = n \cdot_N m = (n \cdot_N m) +_N 0 = (n \cdot_N m) +_N (n \cdot_N 0)$. Next suppose $n \cdot_N (m +_N p) = (n \cdot_N m) +_N (n \cdot_N p)$, for all $n, m \in \mathbb{N}$. Then

$$n \cdot_N (m +_N S(p)) = n \cdot_N S(m +_N p)$$
$$= (n \cdot_N (m +_N p)) +_N n$$
$$= ((n \cdot_N m) +_N (n \cdot_N p)) +_N n$$
$$= (n \cdot_N m) +_N ((n \cdot_N p) +_N n)$$
$$= (n \cdot_N m) +_N (n \cdot_N S(p)),$$

and the theorem follows by induction.

2.13. We prove $(a \cdot_N b) \cdot_N c = a \cdot_N (b \cdot_N c)$ by induction on c:

$$a \cdot_N (b \cdot_N 0) = a \cdot_N 0 = 0 = (a \cdot_N b) \cdot_N 0.$$

If $(a \cdot_N b) \cdot_N c = a \cdot_N (b \cdot_N c)$, then

$$(a \cdot_N b) \cdot_N S(c) = (a \cdot_N b) \cdot_N c +_N (a \cdot_N b)$$
$$= a \cdot_N (b \cdot_N c) +_N (a \cdot_N b)$$
$$= a \cdot_N ((b \cdot_N c) +_N b)$$
$$= a \cdot_N (b \cdot_N S(c)).$$

PROJECT #11. 2.15. First, $0 \cdot_N a = 0$ by induction: $0 \cdot_N 0 = 0$, and if $0 \cdot_N a = 0$, then $0 \cdot_N S(a) = 0 \cdot_N a +_N 0 = 0$.

Now we prove $a \cdot_N b = b \cdot_N a$ by induction on a.

$$0 \cdot_N b = 0 = b \cdot_N 0.$$

Suppose $a \cdot_N b = b \cdot_N a$ for all $b \in \mathbb{N}$. We want to show $S(a) \cdot_N b = b \cdot_N S(a)$. We do this by induction on b: $S(a) \cdot_N 0 = 0 = 0 \cdot_N S(a)$, and if $S(a) \cdot_N b = b \cdot_N S(a)$, then

$$S(a) \cdot_N S(b) = (S(a) \cdot_N b) +_N S(a)$$
$$= (b \cdot_N S(a)) +_N (a +_N 1)$$
$$= (b \cdot_N a) +_N b +_N a +_N 1$$
$$= (a \cdot_N b) +_N a +_N b +_N 1$$
$$= (a \cdot_N S(b)) +_N S(b)$$

$$= (S(b) \cdot_\mathbb{N} a) +_\mathbb{N} S(b)$$
$$= S(b) \cdot_\mathbb{N} S(a).$$

This proves $S(a) \cdot_\mathbb{N} b = b \cdot_\mathbb{N} S(a)$ for all $b \in \mathbb{N}$. This proves $a \cdot_\mathbb{N} b = b \cdot_\mathbb{N} a$ for all $a, b \in \mathbb{N}$.

2.16. $(n +_\mathbb{N} m) \cdot_\mathbb{N} p = p \cdot_\mathbb{N} (n +_\mathbb{N} m) = (p \cdot_\mathbb{N} n) +_\mathbb{N} (p \cdot_\mathbb{N} m) = (n \cdot_\mathbb{N} p) +_\mathbb{N} (m \cdot_\mathbb{N} p)$.

2.17. $2 \cdot_\mathbb{N} 2 = 2 \cdot_\mathbb{N} S(1) = 2 \cdot_\mathbb{N} 1 +_\mathbb{N} 2 = 2 +_\mathbb{N} 2 = 4$.

PROJECT #12. 2.18. Let $\varphi(y)$ be the statement: $\forall x(x <_\mathbb{N} y \to x \in y)$. We use induction on y.

First, notice that $x <_\mathbb{N} 0$ can never happen, since $x <_\mathbb{N} 0$ implies $x +_\mathbb{N} S(k) = 0$, for some $k \in \mathbb{N}$, so $S(x +_\mathbb{N} k) = 0$—impossible. This gives us that $\varphi(0)$ is true.

Now suppose $\varphi(y)$. If $x <_\mathbb{N} S(y)$, then $x +_\mathbb{N} S(k) = S(y)$ for some $k \in \mathbb{N}$, so $S(x +_\mathbb{N} k) = S(y)$, and hence $x +_\mathbb{N} k = y$.

Case 1. $k = 0$. Then $x = y$, and $y \in y \cup \{y\} = S(y)$.
Case 2. $k \neq 0$. Then $k = S(t)$ for some t, so $x <_\mathbb{N} y$ and $x \in y$ (since $\varphi(y)$), and again, $x \in y \cup \{y\} = S(y)$.

This gives us $\varphi(S(y))$, so $\varphi(y)$ is true for all $y \in \mathbb{N}$, by induction.

2.19.

(1) $<_\mathbb{N}$ is irreflexive, since if $a <_\mathbb{N} a$, then $a \in a$, which is impossible.
(2) $<_\mathbb{N}$ is transitive, since if $a <_\mathbb{N} b$ and $b <_\mathbb{N} c$, then $a +_\mathbb{N} S(k) = b$ and $b +_\mathbb{N} S(t) = c$, for some $k, t \in \mathbb{N}$, so $(a +_\mathbb{N} S(k)) +_\mathbb{N} S(t) = c$, so $a +_\mathbb{N} (S(k) +_\mathbb{N} S(t)) = c$, so $a +_\mathbb{N} S(S(k) +_\mathbb{N} t) = c$, so $a <_\mathbb{N} c$.
(3) $<_\mathbb{N}$ satisfies trichotomy: Let $\varphi(b)$ be the statement: $\forall a(a = b \lor a <_\mathbb{N} b \lor b <_\mathbb{N} a)$. $\varphi(0)$ is true, since if $a \neq 0$, then $0 + a = a$, so by definition, $0 <_\mathbb{N} a$. Now suppose $\varphi(b)$ is true.
Case 1. $a = b$. Then $a +_\mathbb{N} 1 = S(b)$, so $a <_\mathbb{N} S(b)$.
Case 2. $a <_\mathbb{N} b$. Then since $b +_\mathbb{N} 1 = S(b)$, $b <_\mathbb{N} S(b)$, so $a <_\mathbb{N} S(b)$.
Case 3. $b <_\mathbb{N} a$. Then $b +_\mathbb{N} S(k) = a$.
    Case 3a. $k = 0$. Then $S(b) = a$.
    Case 3b. $k \neq 0$. Then $k = S(t)$ for some $t \in \mathbb{N}$, so $S(b) +_\mathbb{N} S(t) = S(b +_\mathbb{N} S(t)) = b +_\mathbb{N} S(S(t)) = a$, and so $S(b) <_\mathbb{N} a$.

All these cases imply $\varphi(S(b))$. By induction, $\varphi(b)$ is true for all $b \in \mathbb{N}$.

CHAPTER 3
The Integers

PROJECT #13. 3.1. $\langle a,b \rangle \sim \langle a,b \rangle$ since $a +_{\mathbb{N}} b = b +_{\mathbb{N}} a$. If $\langle a,b \rangle \sim \langle c,d \rangle$ then $a +_{\mathbb{N}} d = b +_{\mathbb{N}} c$, so $c +_{\mathbb{N}} b = d +_{\mathbb{N}} a$, so $\langle c,d \rangle \sim \langle a,b \rangle$. To prove transitivity, we need a lemma we will call:

2.20. Lemma. *For $a, b, c \in \mathbb{N}$: if $a +_{\mathbb{N}} b = a +_{\mathbb{N}} c$, then $b = c$.*

PROOF. By induction on a: $0 +_{\mathbb{N}} b = 0 +_{\mathbb{N}} c$ implies $b = c$. Now suppose that whenever $a +_{\mathbb{N}} b = a +_{\mathbb{N}} c$, then $b = c$. Then if we are given $S(a) +_{\mathbb{N}} b = S(a) +_{\mathbb{N}} c$, then $S(a +_{\mathbb{N}} b) = S(a +_{\mathbb{N}} c)$, so $a +_{\mathbb{N}} b = a +_{\mathbb{N}} c$, so $b = c$. This proves the lemma. □

Now transitivity: if $\langle a,b \rangle \sim \langle c,d \rangle$ and $\langle c,d \rangle \sim \langle e,f \rangle$, then
$$a +_{\mathbb{N}} d = b +_{\mathbb{N}} c \text{ and } c +_{\mathbb{N}} f = d +_{\mathbb{N}} e.$$
Adding:
$$a +_{\mathbb{N}} d +_{\mathbb{N}} c +_{\mathbb{N}} f = b +_{\mathbb{N}} c +_{\mathbb{N}} d +_{\mathbb{N}} e.$$
Cancelling:
$$a +_{\mathbb{N}} f = b +_{\mathbb{N}} e, \quad \text{and so} \quad \langle a,b \rangle \sim \langle e,f \rangle.$$

3.2. $\mathbb{Z}$ is a set by Theorem 1.14.

Defining $+_{\mathbb{Z}}$: We define $[\langle a,b \rangle]_{\sim} +_{\mathbb{Z}} [\langle c,d \rangle]_{\sim} = [\langle a +_{\mathbb{N}} c, b +_{\mathbb{N}} d \rangle]_{\sim}$. We must show (see the suggestions) that this definition doesn't depend on the choice of a, b, c, d. That is, suppose $\langle a,b \rangle \sim \langle a',b' \rangle$ and $\langle c,d \rangle \sim \langle c',d' \rangle$. We must show: $\langle a +_N c, b +_N d \rangle \sim \langle a' +_N c', b' +_N d' \rangle$. Our hypothesis gives us $a +_N b' = b +_N a'$ and $c +_N d' = d +_N c'$. Adding:

$$a +_N b' +_N c +_N d' = b +_N a' +_N d +_N c',$$

which shows:

$$\langle a +_N c, b +_N d \rangle \sim \langle a' +_N c', b' +_N d' \rangle.$$

PROJECT #14.

Definition. Let $0_Z = [\langle 0,0 \rangle]_\sim$.

3.4. $[\langle a,b \rangle]_\sim +_Z 0_Z = [\langle a +_N 0, b +_N 0 \rangle]_\sim = [\langle a,b \rangle]_\sim$.

Definition. For $[\langle c,d \rangle]_\sim \in \mathbb{Z}$, $^-([\langle c,d \rangle]_\sim)_Z = [\langle d,c \rangle]_\sim$.

We must check that if $\langle c,d \rangle \sim \langle c',d' \rangle$, then $\langle d,c \rangle \sim \langle d',c' \rangle$. This is easily done.

3.5. $[\langle c,d \rangle]_\sim +_Z [\langle d,c \rangle]_\sim = [\langle c +_N d, d +_N c \rangle]_\sim$. This equals 0_Z, since $c +_N d +_N 0 = d +_N c +_N 0$.

Definition. $[\langle a,b \rangle]_\sim <_Z [\langle c,d \rangle]_\sim$ iff $a +_N d <_N b +_N c$.

To show this is well-defined, we will need:

2.21. Lemma. *For all $a, b, c \in \mathbb{N}$, $a <_N b$ iff $a +_N c <_N b +_N c$.*

PROOF.

$a <_N b$ iff $a +_N S(k) = b$ for some $k \in \mathbb{N}$
iff $a +_N c +_N S(k) = b +_N c$ for some $k \in \mathbb{N}$
iff $a +_N c <_N b +_N c$. □

Now suppose $\langle a,b \rangle \sim \langle a',b' \rangle$ and $\langle c,d \rangle \sim \langle c',d' \rangle$. Then $a +_N b' = b +_N a'$ and $c +_N d' = d +_N c'$. So,

$a +_N d <_N b +_N c$ iff
$a +_N d +_N a' +_N b' +_N c' +_N d'$
$<_N b +_N c +_N a' +_N b' +_N c' +_N d'$ iff

3. The Integers

$$(a +_N b') +_N (d +_N c') +_N a' +_N d'$$
$$<_N (b +_N a') +_N (c +_N d') +_N b' +_N c' \quad \text{iff}$$
$$a' +_N d' <_N b' +_N c'.$$

3.6. Irreflexive: if $[\langle a,b \rangle]_\sim <_Z [\langle a,b \rangle]_\sim$, then $a +_N b <_N b +_N a$, impossible.

Transitivity: if $[\langle a,b \rangle]_\sim <_Z [\langle c,d \rangle]_\sim$ and $[\langle c,d \rangle]_\sim <_Z [\langle e,f \rangle]_\sim$, then

$$a +_N d <_N b +_N c \text{ and } c +_N f <_N d +_N e,$$

so

$$a +_N d +_N c +_N f <_N b +_N c +_N c +_N f <_N b +_N c +_N d +_N e,$$

so

$$a +_N f <_N b +_N e, \quad \text{and hence} \quad [\langle a,b \rangle]_\sim <_Z [\langle e,f \rangle]_\sim.$$

Trichotomy: given $e = [\langle a,b \rangle]_\sim$, $f = [\langle c,d \rangle]_\sim$, either:

$$a +_N d = b +_N c \text{ (so } e = f),$$

or

$$a +_N d <_N b +_N c \text{ (so } e <_N f),$$

or

$$b +_N c <_N a +_N d \text{ (so } f <_Z e).$$

PROJECT #15.

Definition. $[\langle a,b \rangle]_\sim \cdot_Z [\langle c,d \rangle]_\sim = [\langle (a \cdot_N c) +_N (b \cdot_N d), (b \cdot_N c) +_N (a \cdot_N d) \rangle]_\sim$.

To show this is well-defined, suppose $\langle a,b \rangle \sim \langle a',b' \rangle$ and $\langle c,d \rangle \sim \langle c',d' \rangle$. Then we have

(1) $a +_N b' = a' +_N b$ and
(2) $c +_N d' = c' +_N d$. These give us:
(3) $(a \cdot_N c) +_N (b' \cdot_N c) = (a' \cdot_N c) +_N (b \cdot_N c)$ and
(4) $(b' \cdot_N c) +_N (b' \cdot_N d') = (b' \cdot_N c') +_N (b' \cdot_N d)$, plus
(5) $(a \cdot_N d) +_N (b' \cdot_N d) = (a' \cdot_N d) +_N (b \cdot_N d)$ and
(6) $(a' \cdot_N c) +_N (a' \cdot_N d') = (a' \cdot_N c') +_N (a' \cdot_N d)$. Switching (4) and (5) and then adding all these up, we get:

$$(a \cdot_N c) +_N (b' \cdot_N c) +_N (b' \cdot_N c') +_N (b' \cdot_N d)$$
$$+_N (a' \cdot_N d) +_N (b \cdot_N d) +_N (a' \cdot_N c) +_N (a' \cdot_N d')$$
$$= (a' \cdot_N c) +_N (b \cdot_N c) +_N (b' \cdot_N c) +_N (b' \cdot_N d')$$
$$+_N (a \cdot_N d) +_N (b' \cdot_N d) +_N (a' \cdot_N c') +_N (a' \cdot_N d).$$

With cancellations, we have

$$(a \cdot_N c) +_N (b \cdot_N d) +_N (b' \cdot_N c') +_N (a' \cdot_N d')$$
$$= (a' \cdot_N c') +_N (b' \cdot_N d') +_N (b \cdot_N c) +_N (a \cdot_N d),$$

so

$$\langle (a \cdot_N c) +_N (b \cdot_N d), (b \cdot_N c) +_N (a \cdot_N d) \rangle$$
$$\sim \langle (a' \cdot_N c') +_N (b' \cdot_N d'), (b' \cdot_N c') +_N (a' \cdot_N d') \rangle.$$

3.7. These are routine.

Definition. $1_Z = [\langle 1, 0 \rangle]_\sim$.

3.8. This is routine.

CHAPTER 4
The Rationals

PROJECT #16.

Definition. We define a relation $\approx$ on $\mathbb{Z} \times (\mathbb{Z}\setminus\{0_\mathbb{Z}\})$:

$$\langle a,b \rangle \approx \langle c,d \rangle \quad \text{iff} \quad a \cdot_\mathbb{Z} d = b \cdot_\mathbb{Z} c.$$

This is an equivalence relation: clearly it is reflexive and symmetric. For transitivity, suppose $\langle a,b \rangle \approx \langle c,d \rangle$ and $\langle c,d \rangle \approx \langle e,f \rangle$, so $a \cdot_\mathbb{Z} d = b \cdot_\mathbb{Z} c$ and $c \cdot_\mathbb{Z} f = d \cdot_\mathbb{Z} e$. Multiplying, $a \cdot_\mathbb{Z} d \cdot_\mathbb{Z} f = b \cdot_\mathbb{Z} c \cdot_\mathbb{Z} f$ and $b \cdot_\mathbb{Z} c \cdot_\mathbb{Z} f = b \cdot_\mathbb{Z} d \cdot_\mathbb{Z} e$, so $a \cdot_\mathbb{Z} d \cdot_\mathbb{Z} f = b \cdot_\mathbb{Z} d \cdot_\mathbb{Z} e$. We would like to "cancel" the d, to complete the proof. For this we need first:

2.22. Lemma. (Cancellation for $\cdot_\mathbb{N}$). *For $k, m, n \in \mathbb{N}$, if $S(k) \cdot_\mathbb{N} m = S(k) \cdot_\mathbb{N} n$ then $m = n$.*

PROOF. if not, then either $m <_\mathbb{N} n$ or $n <_\mathbb{N} m$. Suppose $n <_\mathbb{N} m$, so $n +_\mathbb{N} S(t) = m$ for some $t \in \mathbb{N}$. Then

$$S(k) \cdot_\mathbb{N} n = S(k) \cdot_\mathbb{N} n +_\mathbb{N} S(k) \cdot_\mathbb{N} S(t), \quad \text{so}$$

$$0 = S(k) \cdot_\mathbb{N} S(t), \quad \text{so}$$

$$0 = S(k) \cdot_\mathbb{N} t +_\mathbb{N} S(k), \quad \text{so}$$

$$0 = S(S(k) \cdot_\mathbb{N} t +_\mathbb{N} k) \quad \text{which is impossible.} \qquad \square$$

And then we need:

3.9. Lemma (Cancellation for $\cdot_Z$). *For $p, q, r \in \mathbb{Z}$, $q \neq 0_Z$, if $p \cdot_Z q = r \cdot_Z q$ then $p = r$.*

PROOF. Let $p = [\langle i,j \rangle]_\sim$, $q = [\langle k,s \rangle]_\sim$, $r = [\langle t,u \rangle]_\sim$. Since $q \neq 0_Z$, we must have $k \neq s$ (or else $[\langle k,s \rangle]_\sim = [\langle 0,0 \rangle]_\sim$). Now, $p \cdot_Z q = r \cdot_Z q$ means that

$$\langle (i \cdot_N k) +_N (j \cdot_N s), (i \cdot_N s) +_N (j \cdot_N k) \rangle$$
$$\sim \langle (t \cdot_N k) +_N (u \cdot_N s), (t \cdot_N s) +_N (u \cdot_N k) \rangle \text{ so}$$
$$(i \cdot_N k) +_N (j \cdot_N s) +_N (t \cdot_N s) +_N (u \cdot_N k)$$
$$= (i \cdot_N s) +_N (j \cdot_N k) +_N (t \cdot_N k) +_N (u \cdot_N s), \text{ or}$$
$$[(i +_N u) \cdot_N k] +_N [(j +_N t) \cdot_N s]$$
$$= [(i +_N u) \cdot_N s] +_N [(j +_N t) \cdot_N k].$$

Since $k \neq s$, either $k <_N s$ or $s <_N k$. Say for example that $k <_N s$ (the other case is nearly identical). Then $s = k +_N S(v)$ for some v. Substituting and cancelling, we arrive at $(j +_N t) \cdot_N S(v) = (i +_N u) \cdot_N S(v)$. By 2.22, $j +_N t = i +_N u$, so $p = [\langle i,j \rangle]_\sim = [\langle t,u \rangle]_\sim = r$. This proves 3.9 which completes the proof of transitivity. □

Definition. $\mathbb{Q} = \{x \subseteq \mathbb{Z} \times \mathbb{Z} | x = [\langle a,b \rangle]_\approx \text{ for some } b \neq 0_Z\}$.

This is a set by Theorem 1.13.

PROJECT #17.

Definition. $[\langle a,b \rangle]_\approx +_Q [\langle c,d \rangle]_\approx = [\langle (a \cdot_Z d) +_Z (b \cdot_Z c), b \cdot_Z d \rangle]_\approx$.

We must show:

(a) if $b, d \neq 0_Z$, then $b \cdot_Z d \neq 0_Z$, and
(b) $+_Q$ is well-defined.

(a) If $b \cdot_Z d = 0_Z$, then $b \cdot_Z d = 0_Z \cdot_Z d$ (why?), so either $d = 0_Z$, or by Lemma 3.9, $b = 0_Z$.
(b) Suppose $\langle a,b \rangle \approx \langle a',b' \rangle$ and $\langle c,d \rangle \approx \langle c',d' \rangle$. We must show that

$$\langle (a \cdot_Z d) +_Z (b \cdot_Z c), b \cdot_Z d \rangle \approx \langle (a' \cdot_Z d') +_Z (b' \cdot_Z c'), b' \cdot_Z d' \rangle,$$

that is,

4. The Rationals

$$(a \cdot_Z d \cdot_Z b' \cdot_Z d') +_Z (b \cdot_Z c \cdot_Z b' \cdot_Z d')$$
$$= (a' \cdot_Z d' \cdot_Z b \cdot_Z d) +_Z (b' \cdot_Z c' \cdot_Z b \cdot_Z d).$$

Our hypothesis gives us $a \cdot_Z b' = a' \cdot_Z b$ and $c \cdot_Z d' = c' \cdot_Z d$, so $a \cdot_Z b' \cdot_Z d \cdot_Z d' = a' \cdot_Z b \cdot_Z d \cdot_Z d'$ and $b \cdot_Z b' \cdot_Z c \cdot_Z d' = b \cdot_Z b' \cdot_Z c' \cdot_Z d$. Adding these two completes the proof.

4.1. This is straight-forward.

Definition. $0_Q = [\langle 0_Z, 1_Z \rangle]_\approx$.

Definition. $^-([\langle p, q \rangle]_\approx)_Q = [\langle ^-(p)_Z, q \rangle]_\approx$.

This is well-defined, for if $\langle p, q \rangle \approx \langle p', q' \rangle$, then $p \cdot_Z q' = p' \cdot_Z q$, so

$^-(p')_Z \cdot_Z q +_Z {}^-(p)_Z \cdot_Z q' +_Z (p \cdot_Z q')$
$= (p' \cdot_Z q) +_Z {}^-(p')_Z \cdot_Z q +_Z {}^-(p)_Z \cdot_Z q'$, so
$^-(p')_Z \cdot_Z q +_Z ({}^-(p)_Z +_Z p) \cdot_Z q' = ({}^-(p')_Z +_Z p') \cdot_Z q +_Z {}^-(p)_Z \cdot_Z q'$, so
$^-(p')_Z \cdot_Z q +_Z 0_Z = 0_Z +_Z {}^-(p)_Z \cdot_Z q'$, so
$^-(p')_Z \cdot_Z q = {}^-(p)_Z \cdot_Z q'$, so
$\langle ^-(p')_Z, q' \rangle \approx \langle ^-(p)_Z, q \rangle$.

4.2. This is straight-forward.

Project #18.

Definition. $[\langle a, b \rangle]_\approx \cdot_Q [\langle c, d \rangle]_\approx = [\langle a \cdot_Z c, b \cdot_Z d \rangle]_\approx$.

It is routine to show this is well-defined.

4.3. This is tedious, but not difficult.

Definition. $1_Q = [\langle 1_Z, 1_Z \rangle]_\approx$; $(1/[\langle a, b \rangle]_\approx)_Q = [\langle b, a \rangle]_\approx$.

We must show: (a) if $[\langle a, b \rangle]_\approx \neq 0_Q$ then $a \neq 0_Z$, and (b) this is well-defined.

(a) If $a = 0_Z$, then $\langle a, b \rangle \approx \langle 0_Z, 1_Z \rangle$, so $[\langle a, b \rangle]_\approx = 0_Q$.
(b) Routine. (if $\langle a, b \rangle \approx \langle a', b' \rangle$ then $a \cdot_Z b' = a' \cdot_Z b$, so $\langle b, a \rangle \approx \langle b', a' \rangle$)

4.4. This is routine.

In answer to the question posed in the suggestions: The set $\{[\langle s, 1_Z\rangle]_\approx \in \mathbb{Q} | s \in \mathbb{Z}\}$ is a subset of $\mathbb{Q}$ that behaves exactly like $\mathbb{Z}$.

PROJECT #19. Before going further, we will need a series of lemmas about integers:

Definition. $x \in \mathbb{Z}$ is *pos* iff $0_Z <_Z x$. x is *neg* iff $x <_Z 0_Z$.

3.10. Lemma. *The sum of two pos numbers is pos.*

3.11. Lemma. *The sum of two neg numbers is neg.*

3.12. Lemma. x *is pos iff* $^-(x)_Z$ *is neg.*

3.13. Lemma. *The product of two pos numbers is pos.*

3.14. Lemma. *The product of two neg numbers is pos.*

3.15. Lemma. *The product of a pos and a neg is neg.*

4.6. Lemma. *The sum of two positive numbers is positive.*

4.7. Lemma. *The sum of two negative numbers is negative.*

PROOFS. Let $x = [\langle a, b\rangle]_\sim$, $y = [\langle c, d\rangle]_\sim$, $0_Z = [\langle 0, 0\rangle]_\sim$. □

3.10. $0_Z <_Z x$, $0_Z <_Z y$ mean $a <_N b$ and $c <_N d$. By Lemma 2.21 (see the answers to project #14) $a +_N c <_N b +_N d$, so $0_Z <_Z x +_Z y$.

3.11. This is similar.

3.12. $0_Z <_Z x$ iff $a <_N b$ iff $^-(x)_Z = [\langle b, a\rangle]_\sim <_Z 0_Z$.

3.13. If $b <_N a$ and $d <_N c$, then $a = b +_N S(k)$ and $c = d +_N S(t)$ for some $k, t \in \mathbb{N}$. Then

$(a \cdot_N c) +_N (b \cdot_N d)$
$= (b \cdot_N d) +_N (b \cdot_N S(t)) +_N (S(k) \cdot_N d) +_N (S(k) \cdot_N S(t)))$
$\quad +_N (b \cdot_N d)$

4. The Rationals

$$= [(b +_N S(k)) \cdot_N d] +_N [b \cdot_N (d +_N S(t))] +_N [S(k) \cdot_N S(t)]$$
$$= (a \cdot_N d) +_N (b \cdot_N c) +_N S(k) \cdot_N S(t)$$
$$= (a \cdot_N d) +_N (b \cdot_N c) +_N S(S(k) \cdot_N t +_N k),$$

so

$$(a \cdot_N d) +_N (b \cdot_N c) <_N (a \cdot_N c) +_N (b \cdot_N d),$$

so

$$0_Z <_Z [\langle (a \cdot_N c) +_N (b \cdot_N d), (a \cdot_N d) +_N (b \cdot_N c) \rangle]_\sim = x \cdot_Z y.$$

3.14 and 3.15. These are handled similarly.

4.6. Suppose $[\langle a, b \rangle]_\sim$ and $[\langle c, d \rangle]_\sim$ are positive. To see that the sum, $[\langle (a \cdot_Z d) +_Z (b \cdot_Z c), b \cdot_Z d \rangle]_\sim$ is positive, we simply check four cases:

(a) $0_Z <_Z a, b, c, d$
(b) $a, b <_Z 0_Z <_Z c, d$
(c) $c, d <_Z 0_Z <_Z a, b$
(d) $a, b, c, d <_Z 0_Z$.

These are all easily done with the previous lemmas.

4.7. This follows in the same manner.

Now to show positive (on $\mathbb{Q}$) is well-defined: Suppose $\langle a, b \rangle \approx \langle a', b' \rangle$, that is, $a \cdot_Z b' = b \cdot_Z a'$. Then if $0_Z <_Z a \cdot_Z b$ we can show by cases that $0_Z <_Z a' \cdot_Z b'$ too:

Case 1. $0_Z <_Z a, b$. Then since $b' \neq 0_Z$, either
    (a) $a \cdot_Z b' <_Z 0_Z$ so $b' <_Z 0_Z$ and $a' \cdot_Z b <_Z 0_Z$ so $a' <_Z 0_Z$, so $0_Z <_Z a' \cdot_Z b'$, or
    (b) $0_Z <_Z a \cdot_Z b'$ so $0_Z <_Z b'$ and $0_Z <_Z a' \cdot_Z b$ so $0_Z <_Z a'$, so $0_Z <_Z a' \cdot_Z b'$.

Case 2. $a, b <_Z 0_Z$—this is handled similarly.

Thus $0_Z <_Z a \cdot_Z b$ iff $0_Z <_Z a' \cdot_Z b'$.
In the same way, negative is well-defined.

4.5.

(1) Irreflexivity: if $k <_Q k$, then $k +_Q {}^-(k)_Q = 0_Q$ is positive, but $0_Q = [\langle 0_Z, 1_Z \rangle]_\sim$ is not positive.

(2) Transitivity: if $i <_Q j$ and $j <_Q k$, then $j +_Q {}^-(i)_Q$ and $k +_Q {}^-(j)_Q$ are positive. By 4.6, the sum is positive, so $i <_Q k$.
(3) Trichotomy: for $i, j \in Q$ we have three cases:

Case 1. $j +_Q {}^-(i)_Q$ is positive. Then $i <_Q j$.
Case 2. $j +_Q {}^-(i)_Q = 0_Q$. Then $j +_Q {}^-(i)_Q +_Q i = i$, so $j = i$.
Case 3. $j +_Q {}^-(i)_Q$ is not 0_Q or positive, hence it is negative. Then $i +_Q {}^-(j)_Q$ must be positive, because if it were 0_Q or negative, then by 4.7, $[j +_Q {}^-(i)] +_Q [i +_Q {}^-(j)]$ would be negative, but it is 0_Q. Thus $j <_Q i$.

CHAPTER 5
The Real Numbers

PROJECT #20. 5.1. $\mathbb{R}$ is a definable subset of $P(\mathbb{Q})$, hence a set by Comprehension.

Definition. For $r, s \in \mathbb{R}$, $r <_\mathbb{R} s$ iff $r \subsetneq s$.

5.2. Irreflexivity: $r <_\mathbb{R} r$ is clearly false.
Transitivity: $r <_\mathbb{R} s$, $s <_\mathbb{R} t$ mean $r \subsetneq s$ and $s \subsetneq t$, so $r \subsetneq t$ and $r <_\mathbb{R} t$.
Trichotomy: suppose $r \neq s$. Then there is an x in one but not in the other. Say $x \in r \setminus s$. Then for all $y \in s$, $y <_\mathbb{Q} x$ (if not, then $x \in s$ by (1) of the definition of schnitt) and since $y <_\mathbb{Q} x$, $y \in r$, hence $s \subseteq r$. As $s \neq r$ we have $s \subsetneq r$, so $s <_\mathbb{R} r$.

5.3. Let s be an upper bound for X and let $r = \cup X$. r is a schnitt since

(1) $q \in r$, $p <_\mathbb{Q} q$ imply $q \in t \in X$ for some schnitt t, so $p \in t$ and $p \in r$.
(2) $q \in r$ implies $q \in t \in X$ and since t has no greatest element and $t \subseteq r$, q cannot be the greatest element of r.
(3) Since $X \neq \emptyset$ there is a $t \in X$. Since $t \neq \emptyset$ and $t \subseteq r$, $r \neq \emptyset$.
(4) Since $s \neq \mathbb{Q}$ there is a $p \in \mathbb{Q} \setminus s$. p is not in r since $p \in r$ implies $p \in t \in X$ and $t <_\mathbb{R} s$ so $t \subsetneq s$, but $p \notin s$.

r is an upper bound for X since $t \in X$ implies $t \subseteq r$ so $t \leq_\mathbb{R} r$. Finally, suppose u is any other upper bound for X. Then for every $t \in X$, $t \subseteq u$, so that $r = \cup X \subseteq u$, so $r \leq_\mathbb{R} u$.

PROJECT #21. 5.4. To prove $r +_R s$ is a schnitt, we must check the four properties:

1. If $a <_Q b$, $b \in r +_R s$, then $b = x +_Q y$ where $x \in r$, $y \in s$. Let $z = y +_Q a +_Q {}^-(b)_Q$. We have $a = x +_Q z$ and $x \in r$; to complete the proof we need only show $z \in s$. The following is helpful:

4.8. Lemma. *If $x, y \in \mathbb{Q}$ then*:

(a) ${}^-(x +_Q y)_Q = {}^-(x)_Q +_Q {}^-(y)_Q$, and
(b) ${}^-({}^-(x)_Q)_Q = x$.

PROOF.
$$\begin{aligned}{}^-(x +_Q y)_Q &= {}^-(x +_Q y)_Q +_Q (x +_Q y +_Q {}^-(x)_Q +_Q {}^-(y)_Q) \\ &= ({}^-(x +_Q y)_Q +_Q x +_Q y) +_Q {}^-(x)_Q +_Q {}^-(y)_Q \\ &= {}^-(x)_Q +_Q {}^-(y)_Q.\end{aligned}$$
$$\begin{aligned}{}^-({}^-(x)_Q)_Q &= {}^-({}^-(x)_Q)_Q +_Q ({}^-(x)_Q +_Q x) \\ &= {}^-({}^-(x)_Q)_Q +_Q {}^-(x)_Q) +_Q x \\ &= x.\end{aligned}$$
$\square$

Now, $z <_Q y$, since $y +_Q {}^-(z)_Q = y +_Q {}^-(y)_Q +_Q {}^-(a)_Q +_Q b$ (by 4.8) $= b +_Q {}^-(a)_Q$, which is positive (since $a <_Q b$). Since $y \in s$, $z \in s$.

2. Suppose $a \in r +_R s$, $a = x +_Q y$, $x \in r$, $y \in s$. Since r, s have no greatest elements, there are $x' \in r$, $y' \in s$, $x <_Q x'$, $y <_Q y'$. Then $a <_Q x' +_Q y'$ and $x' +_Q y' \in r +_R s$. This last statement requires another:

4.9. Lemma. *If $w, x, y, z \in \mathbb{Q}$ and $w <_Q x$, $y <_Q z$, then $w +_Q y <_Q x +_Q z$.*

PROOF. $x +_Q {}^-(w)_Q$ and $z +_Q {}^-(y)_Q$ are positive therefore the sum is positive by 4.6 (Project #19). $\square$

PROJECT #22. 5.5. Commutativity is clear. Associativity is clear when we notice: $r +_R (s +_R t) = (r +_R s) +_R t = \{x \mid x = a +_Q b +_Q c, a \in r, b \in s, c \in t\}$, by the associativity of $+_Q$.

Definition. Let $0_R = \{x \mid x \in \mathbb{Q} \wedge x <_Q 0_Q\}$.

5. The Real Numbers

0_R easily satisfies properties (1), (3), and (4). For (2), suppose $q \in 0_R$. Consider $p = q \cdot_Q (1/(1_Q +_Q 1_Q)_Q$. $q <_Q 0_Q$ iff $0_Q +_Q {}^-(q)_Q$ is positive. ${}^-(p)_Q$ is also positive, since $q +_Q {}^-(p)_Q +_Q {}^-(p)_Q = 0$, so ${}^-(p)_Q$ can't be negative or 0 (by 4.7). Thus $p <_Q 0_Q$ and $p \in 0_Q$. Finally, $q <_Q p$ since $p +_Q {}^-(q)_Q$ is positive (it can't be negative or zero since $p +_Q {}^-(q)_Q +_Q p +_Q {}^-(q)_Q = p + p + {}^-(q)_Q +_Q {}^-(q)_Q = {}^-(q)_Q$ is positive).

5.6. If $x \in r +_R 0_R$ then $x = a +_Z b$, $a \in r$, $b \in 0_R$, so $x = a +_Q b <_Q a$, so $x \in r$. Thus $r +_R 0_R \leq_R r$. Could r be greater than $r +_R 0_R$? Only if there is a $y \in r \setminus (r +_R 0_R)$.

```
------------------)---------·---)---------------
                r+ᵣ0ᵣ        y   r
```

Since r has no greatest element, there is a $z \in r$, $y <_Q z$. But then $y = z +_Q (y +_Q {}^-(z)_Q)$, $z \in r$, $(y +_Q {}^-(z)_Q) \in 0_Q$, so $y \in r +_R 0_R$, contradicting our assumption.

Project #23.

Note: answers different from the following are also possible!

Definition. For $r \in \mathbb{R}$, ${}^-(r)_R = \{x \in \mathbb{Q} \mid {}^-(x)_Q \in \mathbb{Q} \setminus r$ and for some $y \in \mathbb{Q} \setminus r$ $y <_Q {}^-(x)_Q\}$.

Definition. For $r, s \in \mathbb{R}$, $0_R <_Q r, s$: we define: $r \cdot_R s = \{x \in \mathbb{Q} \mid x = a \cdot_Q b$ for some $a \in r$, $b \in s$, $0_Q <_Q a$, $0_Q <_Q b\} \cup \{x \in \mathbb{Q} \mid x \leq 0_Q\}$. We also define: $0_R \cdot_R r = 0_R$, $r \cdot_R 0_R = 0_R$, $r \cdot_R {}^-(s)_R = {}^-(r \cdot_R s)_R$, ${}^-(r)_R \cdot_R s = {}^-(r \cdot_R s)_R$, and ${}^-(r)_R \cdot_R {}^-(s)_R = r \cdot_R s$.

Definition. $1_R = \{x \mid x <_Q 1_Q\}$.

Definition. For $r \in \mathbb{R}$, $0_R <_R r$, $(1/r)_R = \{x \in \mathbb{Q} \mid \exists y (y \notin r \wedge x <_Q (1/y)_Q)\}$. For $r <_R 0_R$, $(1/r)_R = {}^-((1/{}^-(r)_R)_R)_R$.

Note: if we were continuing, it would be necessary to prove that all of these are schnitts. In the case of $\cdot_R$ we would also have to prove that $t <_R 0_R$ iff $0_R <_R {}^-(t)_R$ to show that the definition given is complete.

The set: $\{\{p \in \mathbb{Q} \mid p <_Q q\} \in \mathbb{R} \mid q \in \mathbb{Q}\}$ is a subset of $\mathbb{R}$ which behaves just like $\mathbb{Q}$.

CHAPTER 6
The Ordinals

PROJECT #24. 6.1. True vacuously.

6.2. To prove Lemma 6.2a: given A, there is a $b \in A$ such that $b \cap A = \emptyset$. This b is $\in$-least in A, that is, if c is any other element of A, then $c \notin b$.

Next, the fact that $a, b \in \mathbb{N}$, $a <_\mathbb{N} b$ iff $a \in b$ is an easy consequence of Theorem 2.18 and the trichotomy property for $<_\mathbb{N}$. This shows parts (1') and (2') of 6.2b. The proof of (3') is an easy induction.

Incidentally, $\{\{\emptyset\}\}$ satisfies (1') vacuously, but fails to satisfy (3'). On the other hand, $\{\emptyset, \{\emptyset\}, \{\{\emptyset\}\}\}$ satisfies (3') but not (1').

6.3. For transitivity, suppose $a, b, c \in S(\alpha) = \alpha \cup \{\alpha\}$ and $a \in b \in c$. Then either (i) all three are in α, so $a \in c$ since α is an ordinal satisfying property (1'); or (ii) $a, b \in \alpha$, $c = \alpha$ and so $a \in b$ by property (3').

For trichotomy, suppose $a, b \in S(\alpha) = \alpha \cup \{\alpha\}$. If $a, b \in \alpha$ we use the trichotomy of $\in$ on α. If $a \in \alpha$, $b = \alpha$ then obviously $a \in b$. Finally, if both a, b equal α, then $a = b$.

For property (3'), suppose $a \in b \in S(\alpha) = \alpha \cup \{\alpha\}$. Then either (i) $b \in \alpha$, so $a \in \alpha$ and then $a \in S(\alpha)$; or (ii) $b = \alpha$, so $a \in \alpha$ and again, $a \in S(\alpha)$.

6.4. First, $b \subseteq \alpha$ since $a \in b \rightarrow a \in \alpha$ by property (3') for α. Since $\in$ is a well-ordering on α, it is a well-ordering on every subset, in particular, b. Second, suppose $c \in d \in b \in \alpha$. Then $c \in d \in \alpha$ and $c \in \alpha$, by property (3') for α. Then since $c, d, b \in \alpha$, we have $c \in b$ by the transitivity of $\in$ on α.

PROJECT #25. 6.5. If $\alpha \in \beta$, then $\gamma \in \alpha \in \beta \rightarrow \gamma \in \beta$ so $\alpha \subseteq \beta$. Since $\alpha \neq \beta$, $\alpha \subsetneq \beta$.

If $\alpha \subsetneq \beta$, let γ be the least element of $\beta \setminus \alpha$. We claim $\alpha = \gamma$ and so $\alpha \in \beta$. Suppose $\delta \in \gamma$. Then $\delta \notin \beta \setminus \alpha$ yet $\delta \in \gamma \in \beta$ so $\delta \in \beta$. This means $\delta \in \alpha$ so we have $\gamma \subseteq \alpha$. To show $\alpha \subseteq \gamma$ and finish the proof, suppose $\delta \in \alpha$. Then $\delta \in \beta$ so by trichotomy either

(1) $\gamma \in \delta$ (then $\gamma \in \alpha$—not true!)
(2) $\gamma = \delta$ (then again $\gamma \in \alpha$—impossible!), or
(3) $\delta \in \gamma$.

This last must be the case, so $\alpha \subseteq \gamma$ and we are done.

6.6. Trans. follows from 1.5. For trich.: suppose α, β are ordinals. If $\alpha \neq \beta$, then one of the two sets $\alpha \setminus \beta$, $\beta \setminus \alpha$ is nonempty. Suppose $\beta \setminus \alpha \neq \emptyset$. Let γ be the least element of $\beta \setminus \alpha$. As in the proof of Theorem 6.5, $\gamma \subseteq \alpha$. Since $\gamma \notin \alpha$ we can't have $\gamma \subsetneq \alpha$ (by Theorem 6.5) and so $\gamma = \alpha$. Thus $\alpha \in \beta$.

6.7. Suppose $A \neq \emptyset$ is a set of ordinals. Let $\alpha \in A$. If $\alpha \cap A = \emptyset$, then α is the least element of A. Otherwise, since $\alpha \cap A \subseteq \alpha$, $\alpha \cap A$ has a least element, β, and this is the least element in A (for $\gamma \in A$ either $\alpha \in \gamma$ so $\beta \in \gamma$, or $\alpha = \gamma$ so $\beta \in \gamma$, or $\gamma \in \alpha$ so $\gamma \in \alpha \cap A$ so $\beta = \gamma$ or $\beta \in \gamma$).

6.8. (1′) and (2′) follow from 6.7. For (3′): if $\alpha \in \beta \in \cup A$, then $\beta \in \gamma$ for some $\gamma \in A$, so $\alpha \in \gamma$, so $\alpha \in \cup A$.

If $\alpha \in A$, then $\alpha \subseteq \cup A$ (so $\alpha = \cup A$ or $\alpha \in \cup A$), so $\cup A$ is an upper bound for A. If γ is any other ordinal such that $\alpha \in A \rightarrow \alpha \subseteq \gamma$, then $\cup A \subseteq \gamma$ (so $\cup A = \gamma$ or $\cup A \in \gamma$), hence $\cup A$ is the least upper bound.

6.9. $\alpha \in S(\alpha)$.

6.10. If the collection of all ordinals were a set then $\cup A$ would be the largest ordinal, which is false by Theorem 6.9.

PROJECT #26. 6.11. ω is not 0, and $\omega \neq S(n)$ for any $n \in \omega$, so it is a limit ordinal. By definition, if $\alpha \in \omega$ then either $\alpha = 0$ or $\alpha = S(n)$ for some $n \in \omega$. Hence α is not a limit ordinal, and so ω is the least such.

6.12. Given φ, suppose there is some γ such that $\varphi(\gamma)$ is not true. There must be a least such ordinal, since

6. The Ordinals

$$X = \{\alpha \in \gamma \,|\, \varphi(\alpha) \text{ is false}\}$$

is either empty (so γ is least) or not (so X has a least member by 6.7). Let β be the least. This, however, contradicts our assumption, since for all $\alpha \in \beta$, $\varphi(\alpha)$ is true.

PROJECT #27. $+_o$ is *not* commutative, $1 +_o \omega = \omega$:

$$\underset{1}{\square} + \underset{\omega}{\square\ \square\ \square} \cdots = \underset{\omega}{\square\ \square\ \square} \cdots$$

but $\omega +_o 1 = S(\omega)$:

$$\underset{\omega}{\square\ \square\ \square} \cdots + \underset{1}{\square} = \underset{S(\omega)}{\square\ \square\ \square \cdots \square}$$

$+_o$ is associative. 0 is the identity. Left cancellation holds, i.e., $\alpha +_o \beta = \alpha +_o \gamma$ implies $\beta = \gamma$, but right cancellation does not, for example, $1 +_o \omega = \omega = 2 +_o \omega$, but $1 \neq 2$.

$\cdot_o$ is *not* commutative, $2 \cdot_o \omega = \omega$:

$$\left[\underset{(\omega)}{(\overset{2}{\square\ \square})(\overset{2}{\square\ \square})(\overset{2}{\square\ \square}) \cdots} \right] = \underset{\omega}{\square\ \square\ \square} \cdots$$

but $\omega \cdot_o 2 = \omega +_o \omega$:

$$\left[\underset{(2)}{(\overset{\omega}{\square\ \square\ \square} \cdots)(\overset{\omega}{\square\ \square\ \square} \cdots)} \right] = \underset{\omega}{\square\ \square} \cdots \underset{+_o}{\ } \underset{\omega}{\square\ \square} \cdots$$

$\cdot_o$ is associative. 1 is the identity. Left cancellation holds, but right cancellation fails, since $1 \cdot_o \omega = 2 \cdot_o \omega$ but $1 \neq 2$.

One distributive law: $\alpha \cdot_o (\beta +_o \gamma) = \alpha \cdot_o \beta +_o \alpha \cdot_o \gamma$ does hold, but the other does not, $(\omega +_o 1) \cdot_o 2 = \omega +_o \omega +_o 1$:

$$\left[\underset{(2)}{(\overset{\omega +_o 1}{\square\ \square \cdots \square})(\overset{\omega +_o 1}{\square\ \square \cdots \square})} \right] = \underset{\omega}{\square\ \square} \cdots \underset{+_o}{\ } \underset{\omega}{\square\ \square} \cdots \underset{+_o\ 1}{\square}$$

but $(\omega \cdot_o 2) +_o (1 \cdot_o 2) = \omega +_o \omega +_o 2$:

$$\left[\underset{(2)}{(\overset{\omega}{\square\ \square} \cdots)(\overset{\omega}{\square\ \square} \cdots)} \right]\left[\underset{(2)}{\overset{1}{\square}\ \overset{1}{\square}} \right]$$

$$= \underset{\omega\ +_o}{\square\ \square} \cdots \underset{\omega\ +_o\ 2}{\square\ \square} \cdots \square\ \square$$

CHAPTER 7
The Cardinals

PROJECT #28. 7.1. For $\|\omega\| \leq \|S(\omega)\|$, the identity map, f, defined by $f(x) = x$, maps ω one-to-one into $S(\omega)$. For $\|S(\omega)\| \leq \|\omega\|$, define f by $f(\omega) = 0$ and $f(n) = n +_{\mathbb{N}} 1$ for all $n \in \omega$.

7.2. For $\|\omega\| \leq \|\mathbb{Z}\|$, let f be the identity map. For $\|\mathbb{Z}\| \leq \|\omega\|$, define f by $f(n) = 2n$, if $0 \leq n$, and $f(-n) = 2n - 1$ if $-n < 0$.

7.3. For $\|\omega\| \leq \|\mathbb{Q}\|$, let f map n to the fraction $n/1$. For $\|\mathbb{Q}\| \leq \|\omega\|$, we make a list. First we list all fractions which can be written using $\{-1, 0, 1\}$:

$$1/1 \qquad 0/1 \qquad -1/1$$

next, all those which can be written with the addition of $\{-2, 2\}$:

$$1/2 \qquad 2/1 \qquad -1/2 \qquad -2/1$$

then all written with the addition of $\{-3, 3\}$:

$$1/3 \quad 2/3 \quad 3/1 \quad 3/2 \quad -1/3 \quad -2/3 \quad -3/1 \quad -3/2$$

and so on. This giant list will contain all the rationals and the length will be exactly ω. This gives us a function:

$$0 \to 1/1$$
$$1 \to 0/1$$

$$2 \to -1/1$$
$$3 \to 1/2$$
$$4 \to 2/1$$
$$\vdots$$

PROJECT #29. 7.4. Following the program of the suggestions, we prove:
(1) $X \subseteq Y \to f(X) \subseteq f(Y) \to B \backslash f(Y) \subseteq B \backslash f(X) \to g(B \backslash f(Y)) \subseteq g(B \backslash f(X)) \to A \backslash g(B \backslash f(X)) \subseteq A \backslash g(B \backslash f(Y))$.
(2) $a \in Z \to a \in X \subseteq H(X)$ for some X. That same X must be entirely contained in Z, so $a \in X \subseteq H(X) \subseteq H(Z)$, so $Z \subseteq H(Z)$.
(3) By definition, $b \in X \subseteq H(X) \subseteq H(Z)$ for some $X \subseteq A$. Then $b \in A \backslash g(B \backslash f(Z))$. Then $a \notin B \backslash f(Z)$—a contradiction.
(4) Since $b \notin H(Z \cup \{b\})$, $b \notin H(Z)$, so $b \notin A \backslash g(B \backslash f(Z))$, so $b \in g(B \backslash f(Z))$, and so $b = g(a)$ for some $a \in B \backslash f(Z)$.
(5) This is clear from (3) and (4) and the fact that f and g are one-to-one.

7.5. True by Theorems 7.4, 7.2 and 7.3.

PROJECT #30. 7.6. Following the suggestions: $C \subseteq A$, so $C \in P(A)$, so $C = f(x)$ for some $x \in A$ since f is onto. We further see that $x \in f(x)$ iff $x \in C$ iff $x \notin f(x)$. This is a contradiction, so our assumption that $\|P(A)\| \leq \|A\|$. is false.

7.7. PROOF #1 given in the suggestions is nearly complete. We only add that the number r constructed is not in the list. It can't be the first number since it differs from the first number in the first digit to the right of the decimal place. It can't be the second, since it differs from the second in the second decimal place, and so on.

Thus the mapping is not onto, a *contradiction*.

PROOF #2 also uses decimal expansions. The mapping takes a set such as $K = \{1, 3, 4, 7, 9, 10, \ldots\}$ to the decimal number:

$$f(K) = .\underset{1}{0}\underset{}{1}\underset{}{0}\underset{3}{1}\underset{4}{1}\underset{}{0}\underset{}{0}\underset{7}{1}\underset{}{0}\underset{9}{1}\underset{}{1} \ldots$$

$$\begin{array}{c} 0\,1\,2\,3\,4\,5\,6\,7\,8\,9 \end{array}$$

It should be clear that if $K_1 \neq K_2$ then $f(K_1) \neq f(K_2)$, since their decimal expansions will be different.

7. The Cardinals

Note that if $\|A\| < \|B\| \le \|C\|$ with f mapping A to B, g mapping B to C one-to-one, then $\|A\| \le \|C\|$ since the composite map $g \circ f$ is one-to-one. On the other hand, $\|A\| < \|C\|$, for if h mapped C to A, one-to-one, then $h \circ g$ would map B to A, one-to-one, contradicting $\|A\| < \|B\|$.

PROOF #3 is complete.

7.8. Let $\alpha = \cup A$. If $\beta \in \alpha$ and $\|\beta\| = \|\alpha\|$, then there is a map f from α to β, one-to-one and onto. $\beta \in \gamma \in A$ for some $\gamma \in A$, so $f \upharpoonright \gamma$ is a one-to-one function from γ into β, contradicting the fact that γ is a cardinal.

7.9. In the suggestions to 7.5, it is shown that $\|\omega\| = \|S(\omega)\|$, hence $S(\omega)$ is not a cardinal.

PROJECT #31. 7.13. By induction on n. 0 is Dedekind finite since it has no proper subsets. Assume that n is Dedekind finite and suppose f maps $n +_\mathbb{N} 1$ to a proper subset of $n +_\mathbb{N} 1$. Let h be as in the suggestions. h is clearly one-to-one, and the range is clearly a subset of n. Furthermore, the range is a proper subset, for there is some $k \le n$ not in the range of f. If $k < n$ then k is not in the range of h, and if $k = n$ then $f(k) < n$ is not in the range of h.

7.14. Let f map n one-to-one onto X, $n \in \mathbb{N}$. Suppose g maps X onto a proper subset of X. Then the function h defined by:

$$h(k) = f^{-1}(g(f(k)))$$

maps n one-to-one onto a proper subset of n (if y is not in the range of g, then $f^{-1}(y)$ is not in the range of h). This contradicts 7.13.

7.15. Use 7.10 to well-order X. By 6.13, there is an ordinal α and a function f mapping α one-to-one onto X. α can't be less than ω, since X is infinite. Then $\omega \le \alpha$, and $f \upharpoonright \omega$ is our required function.

7.16. Let f be the function from the proof of 7.15. Define g from X to X by:

$$g(x) = f(S(f^{-1}(x))).$$

Then g maps X one-to-one into itself. $f(0)$ is not in the range of f, so X is Dedekind infinite.

PROJECT #32. 7.17. Here is a different proof from the one outlined in the Suggestions. Suppose X is a countable set and for each $a \in X$, X_a is a countable set. Let Y be the union of the sets X_a, $Y = \{y | y \in X_a$ for some $a \in X\}$. We must show Y is countable. Let f map X one-to-one into ω, and for each $a \in X$, let f_a map X_a into ω. We need to define a one-to-one map h from Y into ω.

First Attempt. Let $h(y) = (p_n)^{f_a(y)}$ where $y \in X_a$ and $f(a) = n$, and p_n is the nth prime number. The first attempt fails because y might be in two different sets X_a and X_b. Then $h(y)$ would not be well-defined.
Second Attempt. Let $h(y) = (p_n)^{f_a(y)}$ where n is the *smallest* natural number such that for some $a \in X$, $y \in X_a$, and $f(a) = n$. The second attempt succeeds.

The Axiom of Choice was used here to choose the maps f_a. For each a there are infinitely many maps of X_a into ω, but we need a specific one for the proof. In the proof outlined in the suggestions, AC is used to choose the listings.

7.18. For each $a \in A$, let $A_a = A \times \{a\}$. $\|A\| = \|A_a\|$, so each A_a is countable. Then $A \times A = \cup B$ where $B = \{A_a | a \in A\}$ is countable by Theorem 7.17.

7.19. Suppose $\aleph$ is a cardinal. By the Well-ordering Theorem, $P(\aleph)$ is well-orderable. By Theorem 6.13, it is order-isomorphic to an ordinal γ, so $\|P(\aleph)\| = \|\gamma\|$. Let γ be the least such ordinal. By Theorem 7.6, $\|\aleph\| < \|P(\aleph)\|$, so $\aleph \in \gamma$. γ is a cardinal since if $\alpha \in \gamma$ and $\|\alpha\| = \|\gamma\|$, then $\|P(\aleph)\| = \|\alpha\|$, contradicting our choice of γ.

The resolution of Cantor's paradox is that K is not a set.

PROJECT #33. 7.20. Define the ordering $<_x$ on X by: $a <_x b$ iff $f(a) \in f(b)$. It is easy to show this is a linear ordering of X. It is well-ordered since if $A \subseteq X$, $A \neq \emptyset$, then $f(A) \subseteq \beta$ has a least element α. $\alpha = f(a)$ for some $a \in A$, and a is the least element of A. Furthermore, f^{-1} satisfies the conditions in the definition of order-type, so that the order-type of X is α.

7.21. Let $\aleph$ be a cardinal. Let $A = \{\beta | \text{there is a well-ordering of } \aleph \text{ of order-type } \beta\}$. A is a set by Replacement because it is the range of a definable function on a set. The set is the set of well-orderings of $\aleph$ (a definable subset of $P(\aleph \times \aleph)$) and the definable function maps each well-ordering to its order-type. Let $\gamma = \cup A$. γ is greater

7. The Cardinals

than $\aleph$ since there is a well-ordering of $\aleph$ of order-type $\aleph +_o 1$ (define $\alpha < \beta$ iff either $\beta = 0$ or $\alpha \in \beta$, $\alpha \neq 0$). Following the suggestions, γ is a cardinal because if $\alpha < \gamma$, $\|\alpha\| = \|\gamma\|$, then

(1) $\alpha \in \beta \in A$ for some β by definition of γ.
(2) $\|\gamma\| = \|\alpha\| \leqslant \|\beta\| \leqslant \|\gamma\|$ so $\|\gamma\| = \|\beta\|$ by the Shroeder–Bernstein theorem, and hence
(3) there is a one-to-one function from $\aleph$ onto γ.
(4) By 7.20 γ is the order-type of a well-ordering of $\aleph$, so $\gamma \in A$.
(5) If there is a well-ordering of $\aleph$ of length γ, there is one of length $\gamma +_o 1$ as follows: let $T = \aleph \backslash \{0\}$. $\|T\| = \aleph$ so there is a well-ordering of T of length γ. Well-order $\aleph$ by using the ordering on T and putting 0 at the top.
(6) By the above, $\gamma +_o 1 \in A$, so $\gamma +_o 1 \leqslant \cup A = \gamma$—a contradiction.

CHAPTER 8
The Universe

PROJECT #34.

$$V_0 = \{\ \}, \qquad V_1 = \{\{\ \}\}, \qquad V_3 = \{\{\ \},\{\{\ \}\}\},$$
$$V_3 = \{\{\ \},\{\{\ \}\},\{\{\{\ \}\}\},\{\{\ \},\{\{\ \}\}\}\}$$

For free we give you:

$$V_4 = \{\{\ \},\{\{\ \}\},\{\{\{\ \}\}\},\{\{\{\{\ \}\}\}\},\{\{\{\ \},\{\{\ \}\}\}\},$$
$$\{\{\ \},\{\{\ \}\},\{\{\ \},\{\{\{\ \}\}\}\},\{\{\ \},\{\{\ \},\{\{\ \}\}\}\},$$
$$\{\{\{\ \}\},\{\{\{\ \}\}\}\},\{\{\{\ \}\},\{\{\ \},\{\{\ \}\}\}\},\{\{\{\{\ \}\}\},$$
$$\{\{\ \},\{\{\ \}\}\}\},\{\{\ \},\{\{\ \}\},\{\{\{\ \}\}\}\},\{\{\ \},\{\{\ \}\}\},$$
$$\{\{\ \},\{\{\ \}\}\}\},\{\{\ \},\{\{\{\ \}\}\},\{\{\ \},\{\{\ \}\}\}\},\{\{\{\ \}\}\},$$
$$\{\{\{\ \}\}\},\{\{\ \},\{\{\ \}\}\}\},\{\{\ \},\{\{\ \}\},\{\{\{\ \}\}\}\},$$
$$\{\{\ \},\{\{\ \}\}\}\}$$

8.2. For the three cases outlined in the suggestions:

Case 1. Vacuously true, since no $y \in V_0 = 0$.
Case 2. $\alpha = \beta +_o 1$ for some β. Then $y \subseteq V_\beta$, so $x \in V_\beta$. By induction, $z \in x$ implies $z \in V_\beta$, hence $x \subseteq V_\beta$ and so $x \in V_\alpha = P(V_\beta)$.
Case 3. α is a limit ordinal. Then $y \in V_\beta$ for some $\beta \in \alpha$. By induction, $x \in V_\beta$, and so $x \in V_\alpha$.

8.3. Continuing the suggestions:

Case 1. $\alpha = 0$—again, vacuously true.
Case 2. $\alpha = \beta +_o 1$ for some β. Either $\delta = \beta$ or $\delta \in \beta$, but in either situation, $V_\delta \subseteq V_\beta$ using the induction hypothesis. Then if $x \in V_\delta$, $x \in V_\beta$ and then $x \subseteq V_\beta$ by Theorem 8.2, hence $x \in V_\alpha = P(V_\beta)$.
Case 3. α is a limit ordinal. By definition, $V_\delta \subseteq V_\alpha$ for all $\delta \in \alpha$.

PROJECT #35. 8.4. We fill in the gaps left by the suggestions paragraph by paragraph:

(1) $x \subseteq V_\beta$ and $V_{\beta+1} = P(V_\beta)$ so $x \in V_{\beta+1}$.
(2) $\exists f (\forall c \forall d \forall e ((\langle c,d \rangle \in f \wedge \langle c,e \rangle \in f) \to d = e)$

$$\wedge \langle 0, z \rangle \in f$$

$$\wedge \forall c \forall d (\langle c, d \rangle \in f \to \langle S(c), \cup d \rangle \in f)$$

$$\wedge \langle a, b \rangle \in f).$$

(3) If $d \in T(z)$ then $d \in f(n)$ for some n, so $c \in \cup f(n) = f(S(n))$, so $c \in T(z)$.
(4) We can express "$s \in V_\alpha$" (see the proof of Theorem 8.1), so we can express "$\forall \alpha s \notin V_\alpha$".
(5) $q \in x \in T(z)$ implies $q \in T(z)$ by (3). Also, $q \notin Y$, hence q *is* in some V_α.
(6) If $q \in x$, then $q \in V_{g(q)}$ and $g(q) \in R$, so $g(q) \leq \beta$. By Theorem 8.3, $V_{g(q)} \subseteq V_\beta$, so $q \in V_\beta$.

PROJECT #36. 8.5. Following the suggestions:

(1) True by Theorem 7.14.
(2) $\alpha < \omega$ since otherwise X would be Dedekind infinite (see the proof of Theorem 7.16).
(3) $\alpha \neq 0$ since $X \neq 0$, hence $\alpha = S(k)$ for some $k \in \omega$.
(4) Since f is order-preserving, $b < k$ implies $f(b) < f(k)$, so that $f(k)$ is the largest element of X. By Theorem 6.8, $\cup X$ is the least upper bound of X, so $\cup X = f(k)$.
(5) $\cup X = f(k) \in X \subseteq \omega$.

8.6. Let $X = \{\aleph_n | n \in \omega\}$. By definition, $\cup X = \aleph_\omega$ and clearly $\|X\| = \omega$. Note that we can show X is a set by Replacement.

8. The Universe

8.7. Suppose $X \subseteq \aleph_1$, $\|X\| \leq \omega$. Then $\cup X$ is a countable union of countable sets, hence $\|\cup X\| \leq \omega \neq \aleph_1$.

Project #37. 8.8. ω is not strongly inaccessible by definition. $\aleph_\omega$ is not regular. For $\aleph_1$, $X = \aleph_0 = \omega < \aleph_1$, but if $\|P(\aleph_0)\| < \aleph_1$ then $\|P(\aleph_0)\| \leq \aleph_0$, contradicting Theorem 7.6. Similarly for $\aleph_6$, $\|P(\aleph_5)\|$ is not less than $\aleph_6$.

8.9. Let α be the least ordinal such that $\|V_\alpha\|$ is not less than κ. α is clearly not 0.

Case 1. $\alpha = \beta +_o 1$ for some $\beta < \kappa$. Then $\|V_\alpha\| = \|P(V_\beta)\|$ and since $\|V_\beta\| < \kappa$, $\|V_\alpha\| < \kappa$ by the inaccessibility of κ.

Case 2. α is a limit ordinal. Suppose $\kappa \leq \|V_\alpha\|$. Let f map κ one-to-one into V_α. For each $\beta < \alpha$, let $X_\beta = \{\delta \in \kappa \mid f(\delta) \in V_\beta\}$. Then
(1) $\|X_\beta\| \leq \|V_\beta\| < \kappa$,
(2) $\cup X_\beta < \kappa$ since κ is regular,
(3) $K = \cup\{\cup X_\beta \mid \beta < \alpha\} < \kappa$ since κ is regular,
(4) but $K = \kappa$, a contradiction.

Project #38. 8.10. Extension:—obvious.

Empty Set: $\varnothing \in V_1 \subseteq V_\kappa$.

Pair Set: if $x, y \in V_\kappa$, then $x \in V_\alpha$, $y \in V_\beta$ for some $\alpha, \beta < \kappa$. Then if γ is the larger of the two ordinals, $\gamma < \kappa$, $\{x, y\} \subseteq V_\gamma$, so $\{x, y\} \in V_{\gamma+1} \subseteq V_\kappa$.

Union: if $x \in V_\kappa$, then $x \in V_\alpha$ for some $\alpha < \kappa$. If $a \in b \in x$ then $a \in V_\alpha$ by Theorem 8.2, so $\cup x \subseteq V_\alpha$, so $\cup x \in V_{\alpha+1} \subseteq V_\kappa$.

Power Set: if $x \in V_\kappa$, then $x \in V_\alpha$ for some $\alpha < \kappa$. By Theorem 8.2, $x \subseteq V_\alpha$, so $P(x) \subseteq P(V_\alpha) = V_{\alpha+1}$ so $P(x) \in V_{\alpha+2} \subseteq V_\kappa$.

Regularity: —routine, using Theorem 8.2.

Comprehension: if $x \in V_\kappa$ then $x \in V_\alpha$ for some $\alpha < \kappa$. If y is a definable subset of x, then it is a set, and $y \subseteq V_\alpha$, so $y \in V_{\alpha+1} \subseteq V_\kappa$.

Infinity: This follows from:

Lemma. *For all ordinals α, $\alpha \in V_{\alpha+_o 1}$.*

Proof by Transfinite Induction: $0 \in V_1$.

If $\beta \in V_{\beta+1}$, $\beta + 1 = \beta \cup \{\beta\} \subseteq V_{\beta+1}$ so $\beta + 1 \in V_{\beta+2}$.

If λ is a limit ordinal and $\alpha < \lambda$ implies $\alpha \in V_{\alpha+1}$, then $\lambda \subseteq V_\lambda$, so $\lambda \in V_{\lambda+1}$. □

Replacement: suppose f is a function from x to V_κ, $x \in V_\kappa$. Then $x \in V_\alpha$ for some $\alpha < \kappa$. $x \subseteq V_\alpha$ so $\|x\| \leq \|V_\alpha\| < \kappa$, by Theorem 8.9. Define g on x by: $g(a) =$ the least $\beta < \kappa$ such that $f(a) \in V_\beta$. Let R be the range of g. Since $\|x\| < \kappa$, $\|R\| < \kappa$, so $\cup R < \kappa$. As in (6) in the proof of 8.4, paragraph (6), the range of f is contained in $V_{\cup R}$, hence is a member of $V_{\cup R+1} \subseteq V_\kappa$.

CHAPTER 9
Choice and Infinitesimals

PROJECT #39. 9.1. Following the suggestions:
(1) If $\delta < \alpha$ then $g(\alpha) = f(\{g(\beta)|\beta < \alpha\}) = h(X\setminus\{g(\beta)|\beta < \alpha\}) \in X\setminus\{g(\beta)|\beta < \alpha\}$, so $g(\alpha) \neq g(\delta)$.
(2) The range of g is a definable subcollection of X, hence a set.
(3) The domain of g^{-1} is the range of g.
(4) g^{-1} is a function by Replacement; it then follows that g is too.
(5) The domain of g is a collection of ordinals, D. D is an ordinal itself, since if $\beta \in D$ and $\alpha < \beta$, then $\alpha \in D$ by the construction of g.
(6) Let R be the range of f on δ. If $R \neq X$, then $g(\delta) = f(\{g(\beta)|\beta < \alpha\}) = h(X\setminus R)$ is defined, so δ is in the domain of g, so $\delta \in D = \delta$, a contradiction.
(7) g^{-1} maps X one-to-one onto an ordinal, so X is well-ordered by Theorem 7.20.

9.2. (1)–(5) follow as in 9.1.

(6′) g is actually order-preserving: if $\alpha < \beta$, then $g(\alpha) < g(\beta)$, so the range of g, R is well-ordered.
(7′) If α is a limit ordinal, then R has no largest element, so it has a bound not in R. Then $g(\alpha) = f(\{g(\beta)|\beta < \alpha\}) = f(R)$ is defined, so α is in the domain, $\alpha \in \alpha$, a contradiction.
(8′) If $g(\beta)$ were not maximal, then $g(\alpha) = g(\beta +_o 1)$ is defined, again a contradiction.

PROJECT #40. 9.3.

(1) This is routine.
(2) If C is a chain in P, then $\cup C \in P$ is an upper bound for C. $\cup C$ is a function, since if $\langle y, a\rangle, \langle y, b\rangle \in \cup C$, then $\langle y, a\rangle \in f \in C$ and $\langle y, b\rangle \in g \in C$ for some f, g. Since C is linearly ordered by $<_p$, either $f <_p g, f = g$, or $g <_p f$. In any case, $\langle y, a\rangle$ and $\langle y, b\rangle$ both belong to the same function, so a must equal b.
(3) If f is maximal and $y \in X$, $y \neq \emptyset$, then y must be in the domain of f, since otherwise we could choose $a \in y$ and extend f to g:

$$g = f \cup \{\langle y, a\rangle\},$$

then $f <_p g$ and so f is not maximal, a contradiction.

9.4.

(1) This is routine.
(2) The set $\{A | \omega \setminus A \text{ is finite}\}$ is in Q.
(3) If C is a chain in Q, then $\cup C \in Q$ is an upper bound for C. $\cup C$ clearly satisfies (b). Suppose $A, B \in \cup C$. Then $A \in u \in C$, $B \in v \in C$ for some u, v. Since C is linearly ordered by $<_q$, either $u <_q v$, $u = v$, or $v <_q u$. Then either $A, B \in u$ or $A, B \in v$. In both cases $A \cap B \in \cup C$.
(4) Suppose u is a maximal element of Q, $A \subseteq \omega$, and neither A nor $\omega \setminus A$ is in u. We claim that one of the two sets must intersect (must have nonempty intersection with) all members of u.
[PROOF: if not, then for some $B \in u$, $A \cap B = \emptyset$ and for some $D \in u$, $(\omega \setminus A) \cap D = \emptyset$. Then $B \cap D = \emptyset$, so $u \notin Q$, a contradiction.]
Call this set S. S cannot be finite, since then $\omega \setminus S \in u$ and S must intersect all elements of u. Now let $v = \{B \cap S | B \in u\}$. v clearly satisfies (b). It is not difficult to check that v also satisfies (a), so $v \in Q$. Then $u <_q v$, so u is not maximal, a contradiction.

PROJECT #41. 9.5. $\approx$ is clearly reflexive and symmetric. Suppose $f \approx g$ and $g \approx h$, so $X = \{n \in \omega | f(n) = g(n)\}$ and $Y = \{n \in \omega | g(n) = h(n)\}$ are in $\mathscr{U}$. Then $X \cap Y \in \mathscr{U}$ and $X \cap Y \subseteq \{n \in \omega | f(n) = h(n)\}$. It only remains to prove:

9.5a. Lemma. *If $A \in \mathscr{U}$ and $A \subseteq B$, then $B \in \mathscr{U}$.*

PROOF. If $B \notin \mathscr{U}$ then $\omega \setminus B \in \mathscr{U}$, so $A \cap (\omega \setminus B) \in \mathscr{U}$, so $\emptyset \in \mathscr{U}$, contradicting (b) in the definition of ultrafilter. □

9. Choice and Infinitesimals

9.6. $<_{\text{HR}}$ is well-defined: suppose $f \approx f'$ and $g \approx g'$ and $[f]_\approx <_{\text{HR}} [g]_\approx$. Then $A = \{n \in \omega | f(n) = f'(n)\}$, $B = \{n \in \omega | g(n) = g'(n)\}$, and $C = \{n \in \omega | f(n) <_{\mathbb{R}} g(n)\}$ are in $\mathcal{U}$. Hence $A \cap B \cap C \subseteq \{n \in \omega | f'(n) <_{\mathbb{R}} g'(n)\} \in \mathcal{U}$. It is easy to show $<_{\text{HR}}$ is linear.

9.7. Let $H(n) = n$ for all n. Then for any $r \in \mathbb{R}$, if k is an integer greater than r, then $\omega \backslash k \subseteq \{n \in \omega | f_r(n) <_{\text{HR}} H(n)\} \in \mathcal{U}$, so $[f_r]_\approx <_{\text{HR}} [H]_\approx$.

9.8. Let $I(n) = 1/(n + 1)$ for all $n \in \omega$. $\{n \in \omega | I(n) = f_0(n)\} = \emptyset \notin \mathcal{U}$, so $[I]_\approx \neq [f_0]_\approx$. $[I]_\approx$ is certainly positive. Suppose $r \in \mathbb{R}$ is given, $r >_{\mathbb{R}} 0_{\mathbb{R}}$. Then for some k, $1/(k+1) < r$, so $\omega \backslash k \subseteq \{n \in \omega | I(n) < f_r(n)\} \in \mathcal{U}$, so $[I]_\approx$ is infinitesimal.

PROJECT #42. $[f]_\approx +_{\text{HR}} [g]_\approx = [h]_\approx$ where $h(n) = f(n) +_{\mathbb{R}} g(n)$ for all $n \in \omega$. This is well-defined, for if $f \approx f'$ and $g \approx g'$, then

$$\{n \in \omega | f(n) = f'(n)\} \cap \{n \in \omega | g(n) = g'(n)\}$$
$$\subseteq \{n \in \omega | f(n) +_{\mathbb{R}} g(n) = f'(n) +_{\mathbb{R}} g'(n)\}.$$

Similarly, $[f]_\approx \cdot_{\text{HR}} [g]_\approx = [k]_\approx$ where $k(n) = f(n) \cdot_{\mathbb{R}} g(n)$ for all $n \in \omega$ is well-defined.

(1)–(5) are all true. A few proofs (we will use the notation introduced in the suggestions).
  (2) Given any positive $r \in \mathbb{R}$, $\{n \in \omega | I(n) <_{\mathbb{R}} r\} \cap \{n \in \omega | J(n) <_{\mathbb{R}} 1_{\mathbb{R}}\} \subseteq \{n \in \omega | I(n) \cdot_{\mathbb{R}} J(n) <_{\mathbb{R}} r\}$, so $I \cdot_{\text{HR}} J$ is infinitesimal.
  (3) Given any positive $r \in \mathbb{R}$, $\{n \in \omega | I(n) <_{\mathbb{R}} r/2\} \cap \{n \in \omega | J(n) <_{\mathbb{R}} r/2\} \subseteq \{n \in \omega | I(n) +_{\mathbb{R}} J(n) <_{\mathbb{R}} r\}$, so $I +_{\text{HR}} J$ is infinitesimal.
(6)–(8) are all false.
  (8) Suppose $[g]_\approx$ were the smallest infinite, positive hyperreal. Then if $k(n) = g(n) - 1$ for all $n \in \omega$, then $[k]_\approx <_{\text{HR}} [g]_\approx$, but k is also infinite (given $r \in \mathbb{R}$, $\{n \in \omega | k(n) >_{\mathbb{R}} r\} = \{n \in \omega | g(n) >_{\mathbb{R}} r +_{\mathbb{R}} 1_{\mathbb{R}}\} \in \mathcal{U}$).
  (9) True by Theorem 9.9. "$\forall x(\sin^2(x) + \cos^2(x) = 1)$" can be expressed in $\mathscr{L}_{\mathbb{R}}$.
  (10) False. This cannot be expressed in $\mathscr{L}_{\mathbb{R}}$, so Theorem 9.9 doesn't apply (we can't describe "subset", for example). As an example, consider the set of all finite numbers. It has no least upper bound (since the answers to (6) and (8) are both False). Note that this set also cannot be expressed in $\mathscr{L}_{\mathbb{R}}$.

CHAPTER 10
Goodstein's Theorem

PROJECT #43.

(1) $g_2(11) = 84$, $g_3(11) = 1027$, then 15627, 279937, 5764801, 134217727, 2749609302.
(2) $g_2(3) = 3$, then 3, 2, 1, and $g_6(3) = 0$.
(3) First, the answers to some warm-ups: (a) 13 (b) 38 (c) $46 + 92 = 138$.

Now after 1 step:	$2 \cdot 3^2 + 2 \cdot 3 + 2$,
after $1 + 3$ steps:	$2 \cdot 6^2 + 1 \cdot 6 + 5$,
after $1 + 3 + 6$ steps:	$2 \cdot 12^2 + 11$,
after $1 + 3 + 6 + 12$ steps:	$24^2 + 23 \cdot 24 + 23$,
after $1 + 3 + 6 + 12 + 24$ steps:	$48^2 + 22 \cdot 48 + 47$.

Each time we double the increment number of steps, we reduce the middle coefficient by 1. To get rid of the squared term, we will have to do this 23 more times, so we will have proceeded

$$n = 1 + 3 + 6 + 12 + \cdots + 3 \cdot 2^{26} \text{ steps,}$$

and we will have:

$$g_n(4) = (n + 1)(n + 2) + (n + 1) \quad \text{(in base } n + 2\text{)}.$$

To reduce the coefficient of $(n + 2)$ to zero, we must continue the doubling process $n + 1$ more times, so we will have proceeded

$$m = 1 + 3 + 6 + 12 + \cdots + 3 \cdot 2^{26+(n+1)} \text{ steps,}$$

and we will have: $g_m(4) = m + 1$, so the answer is $m + (m + 1)$ steps, or $g_{2m+1}(4) = 0$. We can compute:

$$\begin{aligned} n &= 1 + 3(1 + 2 + 4 + \cdots + 2^{26}) \\ &= 1 + 3(2^{27} - 1) \\ &= 3 \cdot 2^{27} - 2. \end{aligned}$$

Then

$$\begin{aligned} m &= 1 + 3(1 + 2 + 4 + \cdots + 2^{25+3 \cdot 2^{27}}) \\ &= 1 + 3(2^{26+3 \cdot 2^{27}} - 1) \end{aligned}$$

so

$$\begin{aligned} 2m + 1 &= 2 + 3(2^{27+3 \cdot 2^{27}} - 2) + 1 \\ &= 3 \cdot 2^{402653211} - 3, \quad \text{a very large number.} \end{aligned}$$

PROJECT #44. 10.2.

(1) $f_{n+1}(S_n(0)) = f_{n+1}(0) = 0 = f_n(0)$, and

(2) If $m = \sum_{i=0}^{d} k_i \cdot n^i$,

with each $0 \leq k_i < n$, and if the lemma is true for all numbers less than m, then

$$\begin{aligned} f_{n+1}(S_n(m)) &= f_{n+1}\left(S_n\left(\sum_{i=0}^{d} k_i \cdot n^i\right)\right) \\ &= f_{n+1}\left(\sum_{i=0}^{d} S_n(k_i \cdot n^i)\right) \\ &= f_{n+1}\left(\sum_{i=0}^{d} k_i \cdot (n+1)^{S_n(i)}\right) \\ &= \sum_{i=0}^{d} f_{n+1}(k_i \cdot (n+1)^{S_n(i)}) \\ &= \sum_{i=0}^{d} \omega^{f_{n+1}(S_n(i))} \cdot k_i \end{aligned}$$

10. Goodstein's Theorem

$$= \sum_{i=0}^{d} \omega^{f_n(i)} \cdot k_i$$
$$= \sum_{i=0}^{d} f_n(k_i \cdot n^i)$$
$$= f_n\left(\sum_{i=0}^{d} k_i \cdot n^i\right)$$
$$= f_n(m).$$

10.3. Case 1 as outlined in the suggestions is easy, if
$$f_n(m) = \omega^d \cdot k_d + \omega^{d-1} \cdot k_{d-1} \cdots + \omega \cdot k_1 + k_0,$$
then
$$f_n(m+1) = \omega^d \cdot k_d + \omega^{d-1} \cdot k_{d-1} + \cdots + \omega \cdot k_1 + k_0 + 1.$$

For case 2,
$$f_n(k_s \cdot n^s + (n-1) \cdot n^{s-1} + \cdots + (n-1))$$
$$= \omega^s \cdot k_s + \omega^{s-1} \cdot (n-1) + \cdots + (n-1)$$
$$\leqslant \omega^s \cdot k_s + \omega^{s-1} \cdot (n-1) + \cdots + \omega^{s-1} \cdot (n-1)$$
$$\leqslant \omega^s \cdot k_s + \omega^{s-1} \cdot ((n-1) + \cdots + (n-1))$$
$$\leqslant \omega^s \cdot k_s + \omega^{s-1} \cdot (n-1) \cdot s$$
$$< \omega^s \cdot k_s + \omega^s$$
$$= \omega^s \cdot (k_s + 1)$$
$$= f_n((k_s + 1) \cdot n^s).$$

10.4.
$$f_{n+2}(g_{n+1}(m)) = f_{n+2}(S_{n+1}(g_n(m)) - 1)$$
$$< f_{n+2}(S_{n+1}(g_n(m)))$$
$$= f_{n+1}(g_n(m)).$$

10.1. If for some m the sequence: $g_2(m), g_3(m), \ldots$ went on forever, then we would have an infinite descending sequence of ordinals: $f_3(g_2(m)) > f_4(g_3(m)) > f_5(g_4(m)) > \cdots$ which is impossible.

Index

A
AC (*see* Choice, Axiom of)
Associative law
  of addition
    for integers, 22
    for natural numbers, 17, 101–102
    for ordinals, 30, 121
    for rationals, 23
    for reals, 26, 116
  of multiplication
    for integers, 22
    for natural numbers, 18, 55
    for ordinals, 30, 121
    for rationals, 23
    for reals, 26, 116
Axiom of Choice (*see* Choice, Axiom of)

B
Balzac, H., 23
Bounded chain, 34
Bruno, G., 15

C
Cancellation law
  for addition, 57, 105
  for multiplication, 109–110
  for ordinal arithmetic, 30, 121
Cantor, G., 2, 34, 74
Cantor's diagonal proof, 74
Cardinal, 33–36
  regular, 38
  singular, 38
Cartesian product, 12
CH (*see* Continuum Hypothesis)
Chain, 34
Choice, Axiom of, 34
  equivalence with Zermelo's Theorem and Zorn's Lemma, 41–42, 85–87, 133–134
  function, 34
  independence of, 35
Cohen, D. W., v, vi
Cohen, P., 35, 36
Commutative law
  of addition

Commutative law (*cont.*)
   for integers, 22
   for natural numbers, 17, 102
   for ordinals, 30, 121
   for rationals, 23
  of multiplication
   for integers, 22
   for natural numbers, 18, 103–104
   for ordinals, 30, 121
   for rationals, 23
   for reals, 26
Completeness Theorem, 39
Comprehension, Axiom of, 11
Consistency, relative (*see* Relative consistency)
Consistent theory, 29
Con(*T*), 39
Continuity of the reals, 26
Continuum Hypothesis, 36
  Generalized, 36
Cummings, E. E., 37

D

Darwin, C., 27
Dedekind
  finite, 35
  infinite, 35
Distributive law
  for integers, 22
  for natural numbers, 18, 103
  for ordinals, 30, 121
  for rationals, 23
Domain (*see* Function, domain of)

E

Empty set, Axiom of, 10
Equivalence
  class, 14
  relation (*see* Relation, equivalence)

Euclid, 1
Extension, Axiom of, 10

F

Finite
  Dedekind (*see* Dedekind finite)
  hyperreal number, 42
  set, 35
Formalism 2, 4, 29
Fraenkel, A., 2
Function
  domain of, 12
  inverse, 12
  one-to-one, 12
  onto, 12
  restriction of, 12

G

Gamov, G., 71
GCH (*see* Continuum Hypothesis, Generalized)
Goodstein, R. L., 45, 47, 137–139
Goodstein's Theorem, 45–48, 126–127

H

Hartog's Theorem, 36
Hyperreal numbers, 42

I

Identity
  additive
   for integers, 106
   for natural numbers, 17
   for ordinals, 30, 121
   for rationals, 111
   for reals, 116

Index

multiplicative
  for integers, 108
  for natural numbers, 18
  for ordinals, 30, 121
  for rationals, 111
Incompleteness Theorem
  Gödel's first, 35
  Gödel's second, 29, 39, 48
Induction, 16
  double, 56
  transfinite, 28
Infinite
  Dedekind (see Dedekind infinite)
  hyperreal number, 42
  set, 35
Infinitesimal, 43
Infinity, Axiom of, 15
Integer, 21
Intersection of sets, 9, 11
Inverse
  additive
    in the integers, 58
    in the rationals, 60
    in the real numbers, 117
  function (see Function, inverse)
  multiplicative
    in the rationals, 111
Irreflexive ordering (see Ordering, irreflexive)

K
Kirby, L., 45, 47, 48
Kronecker, L., 21, 58, 64

L
$\mathscr{L}$, 7–11
$\mathscr{L}^+$, 10, 11
Least upper bound, 26
Limit ordinal (see Ordinal, limit)

Linear ordering (see Ordering, linear)

M
Maximal element, 34
Model, 39
Moore, R. L., v

N
Natural Number, 16
Neg, 60
Negative
  Integer, 60
  Rational, 24

O
Ordered pair, 12, 53, 98
Ordering
  irreflexive, 18
  linear, 18
  on integers, 22, 107
  on natural numbers, 18, 104
  on ordinals, 28, 120
  on rationals, 24, 112–114
  on reals, 26, 115
  partial, 18
  well- (see Well-ordering)
Order-preserving, 30
Order-type, 30
Ordinal number, 27
  limit, 28
  successor, 28

P
Pair Set, Axiom of, 10
Paradox
  Burali-Forti's, 68

Paradox (*cont.*)
  Cantor's, 79
    resolution of, 126
  Russell's, 2, 3, 53
  Paris, J., 45, 47, 48
Partial ordering (*see* Ordering, partial)
Peano's Axioms, 16, 48
Platonism, 2, 4, 29
Pos, 60
Positive
  Integers, 60
  Rationals, 24
Power set
  axiom of, 10
  existence, 10
  notation, 12
  uniqueness, 11

R
Rational number, 23, 59–60, 110
Real number, 25
Recursive Definitions, Theorem on, 31
Reflexive relation (*see* Relation, reflexive)
Regular cardinal (*see* Cardinal, regrular
Regularity, Axiom of, 10, 82
Relation, 13
  equivalence, 13
  reflexive, 13
  symmetric, 13
  transitive, 13
Relative consistency, 40
Replacement, Axiom of, 28
Russell, B., 2
Russell's Paradox (*see* Paradox, Russell's)

S
Schnitt, 25
Set
  transitive, 68
  well-ordered, 27
Shakespeare, W., 33
Shröder–Bernstein Theorem, 34, 72–74, 124
Singular cardinal (*see* Cardinal, singular)
Strongly inaccessible cardinal, 38
Subset, 9
  proper, 9
Successor ordinal (*see* Ordinal, successor)
Superbase, 45–46
Symmetric relation (*see* Relation, symmetric)

T
Tennyson, A. L., 45
Theory, 39
Thoreau, H. D., 7
Transitive
  closure, 81–82
  relation (*see* Relation, transitive)
  set (*see* Set, transitive)
Trichotomy, 18

U
Ultrafilter, 42
Union
  Axiom of, 10
  of a set, 11, 12
  of two sets, 9, 11
Upper bound, 26

V
von Neumann, J., 25

W

Well-defined, 57–58
Well-ordering, 27
   Theorem (*see* Zermelo's Theorem)
Whitehead, A. N., 41

Z

Zermelo, E., 2
Zermelo–Fraenkel set theory, 2–4
   axioms of, 10–11, 15, 28
   consistency of, 29
   model for, 39
   with Choice, 34
Zermelo's Theorem, 34, 41, 85–86, 133 (*see also* Choice, Axiom of)
ZF (*see* Zermelo–Fraenkel set theory
ZFC, 34
Zorn's Lemma, 34, 41, 42, 85–86, 133–134 (*see also* Choice, Axiom of)

CATALOG OF DOVER BOOKS

Mathematics

FUNCTIONAL ANALYSIS (Second Corrected Edition), George Bachman and Lawrence Narici. Excellent treatment of subject geared toward students with background in linear algebra, advanced calculus, physics and engineering. Text covers introduction to inner-product spaces, normed, metric spaces, and topological spaces; complete orthonormal sets, the Hahn-Banach Theorem and its consequences, and many other related subjects. 1966 ed. 544pp. 6⅛ x 9¼. 0-486-40251-7

ASYMPTOTIC EXPANSIONS OF INTEGRALS, Norman Bleistein & Richard A. Handelsman. Best introduction to important field with applications in a variety of scientific disciplines. New preface. Problems. Diagrams. Tables. Bibliography. Index. 448pp. 5⅜ x 8½. 0-486-65082-0

VECTOR AND TENSOR ANALYSIS WITH APPLICATIONS, A. I. Borisenko and I. E. Tarapov. Concise introduction. Worked-out problems, solutions, exercises. 257pp. 5⅜ x 8¼. 0-486-63833-2

AN INTRODUCTION TO ORDINARY DIFFERENTIAL EQUATIONS, Earl A. Coddington. A thorough and systematic first course in elementary differential equations for undergraduates in mathematics and science, with many exercises and problems (with answers). Index. 304pp. 5⅜ x 8½. 0-486-65942-9

FOURIER SERIES AND ORTHOGONAL FUNCTIONS, Harry F. Davis. An incisive text combining theory and practical example to introduce Fourier series, orthogonal functions and applications of the Fourier method to boundary-value problems. 570 exercises. Answers and notes. 416pp. 5⅜ x 8½. 0-486-65973-9

COMPUTABILITY AND UNSOLVABILITY, Martin Davis. Classic graduate-level introduction to theory of computability, usually referred to as theory of recurrent functions. New preface and appendix. 288pp. 5⅜ x 8½. 0-486-61471-9

ASYMPTOTIC METHODS IN ANALYSIS, N. G. de Bruijn. An inexpensive, comprehensive guide to asymptotic methods–the pioneering work that teaches by explaining worked examples in detail. Index. 224pp. 5⅜ x 8½ 0-486-64221-6

APPLIED COMPLEX VARIABLES, John W. Dettman. Step-by-step coverage of fundamentals of analytic function theory–plus lucid exposition of five important applications: Potential Theory; Ordinary Differential Equations; Fourier Transforms; Laplace Transforms; Asymptotic Expansions. 66 figures. Exercises at chapter ends. 512pp. 5⅜ x 8½. 0-486-64670-X

INTRODUCTION TO LINEAR ALGEBRA AND DIFFERENTIAL EQUATIONS, John W. Dettman. Excellent text covers complex numbers, determinants, orthonormal bases, Laplace transforms, much more. Exercises with solutions. Undergraduate level. 416pp. 5⅜ x 8½. 0-486-65191-6

RIEMANN'S ZETA FUNCTION, H. M. Edwards. Superb, high-level study of landmark 1859 publication entitled "On the Number of Primes Less Than a Given Magnitude" traces developments in mathematical theory that it inspired. xiv+315pp. 5⅜ x 8½. 0-486-41740-9

CATALOG OF DOVER BOOKS

CALCULUS OF VARIATIONS WITH APPLICATIONS, George M. Ewing. Applications-oriented introduction to variational theory develops insight and promotes understanding of specialized books, research papers. Suitable for advanced undergraduate/graduate students as primary, supplementary text. 352pp. 5⅜ x 8½.
0-486-64856-7

COMPLEX VARIABLES, Francis J. Flanigan. Unusual approach, delaying complex algebra till harmonic functions have been analyzed from real variable viewpoint. Includes problems with answers. 364pp. 5⅜ x 8½.
0-486-61388-7

AN INTRODUCTION TO THE CALCULUS OF VARIATIONS, Charles Fox. Graduate-level text covers variations of an integral, isoperimetrical problems, least action, special relativity, approximations, more. References. 279pp. 5⅜ x 8½.
0-486-65499-0

COUNTEREXAMPLES IN ANALYSIS, Bernard R. Gelbaum and John M. H. Olmsted. These counterexamples deal mostly with the part of analysis known as "real variables." The first half covers the real number system, and the second half encompasses higher dimensions. 1962 edition. xxiv+198pp. 5⅜ x 8½. 0-486-42875-3

CATASTROPHE THEORY FOR SCIENTISTS AND ENGINEERS, Robert Gilmore. Advanced-level treatment describes mathematics of theory grounded in the work of Poincaré, R. Thom, other mathematicians. Also important applications to problems in mathematics, physics, chemistry and engineering. 1981 edition. References. 28 tables. 397 black-and-white illustrations. xvii + 666pp. 6⅛ x 9¼.
0-486-67539-4

INTRODUCTION TO DIFFERENCE EQUATIONS, Samuel Goldberg. Exceptionally clear exposition of important discipline with applications to sociology, psychology, economics. Many illustrative examples; over 250 problems. 260pp. 5⅜ x 8½.
0-486-65084-7

NUMERICAL METHODS FOR SCIENTISTS AND ENGINEERS, Richard Hamming. Classic text stresses frequency approach in coverage of algorithms, polynomial approximation, Fourier approximation, exponential approximation, other topics. Revised and enlarged 2nd edition. 721pp. 5⅜ x 8½. 0-486-65241-6

INTRODUCTION TO NUMERICAL ANALYSIS (2nd Edition), F. B. Hildebrand. Classic, fundamental treatment covers computation, approximation, interpolation, numerical differentiation and integration, other topics. 150 new problems. 669pp. 5⅜ x 8½.
0-486-65363-3

THREE PEARLS OF NUMBER THEORY, A. Y. Khinchin. Three compelling puzzles require proof of a basic law governing the world of numbers. Challenges concern van der Waerden's theorem, the Landau-Schnirelmann hypothesis and Mann's theorem, and a solution to Waring's problem. Solutions included. 64pp. 5⅜ x 8½.
0-486-40026-3

THE PHILOSOPHY OF MATHEMATICS: AN INTRODUCTORY ESSAY, Stephan Körner. Surveys the views of Plato, Aristotle, Leibniz & Kant concerning propositions and theories of applied and pure mathematics. Introduction. Two appendices. Index. 198pp. 5⅜ x 8½.
0-486-25048-2

CATALOG OF DOVER BOOKS

INTRODUCTORY REAL ANALYSIS, A.N. Kolmogorov, S. V. Fomin. Translated by Richard A. Silverman. Self-contained, evenly paced introduction to real and functional analysis. Some 350 problems. 403pp. 5⅜ x 8½. 0-486-61226-0

APPLIED ANALYSIS, Cornelius Lanczos. Classic work on analysis and design of finite processes for approximating solution of analytical problems. Algebraic equations, matrices, harmonic analysis, quadrature methods, much more. 559pp. 5⅜ x 8½. 0-486-65656-X

AN INTRODUCTION TO ALGEBRAIC STRUCTURES, Joseph Landin. Superb self-contained text covers "abstract algebra": sets and numbers, theory of groups, theory of rings, much more. Numerous well-chosen examples, exercises. 247pp. 5⅜ x 8½. 0-486-65940-2

QUALITATIVE THEORY OF DIFFERENTIAL EQUATIONS, V. V. Nemytskii and V.V. Stepanov. Classic graduate-level text by two prominent Soviet mathematicians covers classical differential equations as well as topological dynamics and ergodic theory. Bibliographies. 523pp. 5⅜ x 8½. 0-486-65954-2

THEORY OF MATRICES, Sam Perlis. Outstanding text covering rank, nonsingularity and inverses in connection with the development of canonical matrices under the relation of equivalence, and without the intervention of determinants. Includes exercises. 237pp. 5⅜ x 8½. 0-486-66810-X

INTRODUCTION TO ANALYSIS, Maxwell Rosenlicht. Unusually clear, accessible coverage of set theory, real number system, metric spaces, continuous functions, Riemann integration, multiple integrals, more. Wide range of problems. Undergraduate level. Bibliography. 254pp. 5⅜ x 8½. 0-486-65038-3

MODERN NONLINEAR EQUATIONS, Thomas L. Saaty. Emphasizes practical solution of problems; covers seven types of equations. ". . . a welcome contribution to the existing literature...."—*Math Reviews.* 490pp. 5⅜ x 8½. 0-486-64232-1

MATRICES AND LINEAR ALGEBRA, Hans Schneider and George Phillip Barker. Basic textbook covers theory of matrices and its applications to systems of linear equations and related topics such as determinants, eigenvalues and differential equations. Numerous exercises. 432pp. 5⅜ x 8½. 0-486-66014-1

LINEAR ALGEBRA, Georgi E. Shilov. Determinants, linear spaces, matrix algebras, similar topics. For advanced undergraduates, graduates. Silverman translation. 387pp. 5⅜ x 8½. 0-486-63518-X

ELEMENTS OF REAL ANALYSIS, David A. Sprecher. Classic text covers fundamental concepts, real number system, point sets, functions of a real variable, Fourier series, much more. Over 500 exercises. 352pp. 5⅜ x 8½. 0-486-65385-4

SET THEORY AND LOGIC, Robert R. Stoll. Lucid introduction to unified theory of mathematical concepts. Set theory and logic seen as tools for conceptual understanding of real number system. 496pp. 5⅜ x 8¼. 0-486-63829-4

CATALOG OF DOVER BOOKS

TENSOR CALCULUS, J.L. Synge and A. Schild. Widely used introductory text covers spaces and tensors, basic operations in Riemannian space, non-Riemannian spaces, etc. 324pp. 5⅝ x 8¼. 0-486-63612-7

ORDINARY DIFFERENTIAL EQUATIONS, Morris Tenenbaum and Harry Pollard. Exhaustive survey of ordinary differential equations for undergraduates in mathematics, engineering, science. Thorough analysis of theorems. Diagrams. Bibliography. Index. 818pp. 5⅜ x 8½. 0-486-64940-7

INTEGRAL EQUATIONS, F. G. Tricomi. Authoritative, well-written treatment of extremely useful mathematical tool with wide applications. Volterra Equations, Fredholm Equations, much more. Advanced undergraduate to graduate level. Exercises. Bibliography. 238pp. 5⅜ x 8½. 0-486-64828-1

FOURIER SERIES, Georgi P. Tolstov. Translated by Richard A. Silverman. A valuable addition to the literature on the subject, moving clearly from subject to subject and theorem to theorem. 107 problems, answers. 336pp. 5⅜ x 8½. 0-486-63317-9

INTRODUCTION TO MATHEMATICAL THINKING, Friedrich Waismann. Examinations of arithmetic, geometry, and theory of integers; rational and natural numbers; complete induction; limit and point of accumulation; remarkable curves; complex and hypercomplex numbers, more. 1959 ed. 27 figures. xii+260pp. 5⅜ x 8½. 0-486-63317-9

POPULAR LECTURES ON MATHEMATICAL LOGIC, Hao Wang. Noted logician's lucid treatment of historical developments, set theory, model theory, recursion theory and constructivism, proof theory, more. 3 appendixes. Bibliography. 1981 edition. ix + 283pp. 5⅜ x 8½. 0-486-67632-3

CALCULUS OF VARIATIONS, Robert Weinstock. Basic introduction covering isoperimetric problems, theory of elasticity, quantum mechanics, electrostatics, etc. Exercises throughout. 326pp. 5⅜ x 8½. 0-486-63069-2

THE CONTINUUM: A CRITICAL EXAMINATION OF THE FOUNDATION OF ANALYSIS, Hermann Weyl. Classic of 20th-century foundational research deals with the conceptual problem posed by the continuum. 156pp. 5⅜ x 8½. 0-486-67982-9

CHALLENGING MATHEMATICAL PROBLEMS WITH ELEMENTARY SOLUTIONS, A. M. Yaglom and I. M. Yaglom. Over 170 challenging problems on probability theory, combinatorial analysis, points and lines, topology, convex polygons, many other topics. Solutions. Total of 445pp. 5⅜ x 8½. Two-vol. set.
Vol. I: 0-486-65536-9 Vol. II: 0-486-65537-7

Paperbound unless otherwise indicated. Available at your book dealer, online at **www.doverpublications.com**, or by writing to Dept. GI, Dover Publications, Inc., 31 East 2nd Street, Mineola, NY 11501. For current price information or for free catalogues (please indicate field of interest), write to Dover Publications or log on to **www.doverpublications.com** and see every Dover book in print. Dover publishes more than 500 books each year on science, elementary and advanced mathematics, biology, music, art, literary history, social sciences, and other areas.